Transforming Outcome-Based Education with Machine Learning

D. Vetrithangam • Puneet Kumar
Jaspreet Singh Batth • B. Arunadevi
V. Saravanan
Editors

Transforming Outcome-Based Education with Machine Learning

Springer

Editors
D. Vetrithangam
Department of Computer Science & Engineering
University Institute of Engineering
Chandigarh University
Mohali, Punjab, India

Jaspreet Singh Batth
Department of Computer Science & Engineering
Chandigarh University
Mohali, Punjab, India

V. Saravanan
Dambi Dollo University
Dembi Dolo, Ethiopia

Puneet Kumar
Department of Computer Science & Engineering
Chandigarh University
Mohali, Punjab, India

B. Arunadevi
Department of Electronics and Communication Engineering
Dr.N.G.P Institute of Technology
Coimbatore, Tamil Nadu, India

ISBN 978-981-95-4924-5 ISBN 978-981-95-4925-2 (eBook)
https://doi.org/10.1007/978-981-95-4925-2

This Springer imprint is published by the registered company Springer Nature Singapore Pte Ltd.
The registered company address is: 152 Beach Road, #21-01/04 Gateway East, Singapore 189721, Singapore

Contents

Genesis and Growth of Outcome-Based Education in Higher Education

D. Vetrithangam, Puneet Kumar, and Jaspreet Singh Batth

Abstract This chapter describes the evolution of Outcome-Based Education (OBE) and its significant role in transforming the structure and practices of higher education systems. It begins by comparing traditional teaching methods—focused on lectures, fixed schedules, and memorization—with a more student-centered approach that emphasizes practical skills and measurable outcomes. This shift laid the foundation for Competency-Based Education (CBE). The chapter explains how OBE aligns with international education standards and meets the accreditation bodies' requirements, thereby ensuring its global relevance. It also outlines key milestones in the worldwide adoption of OBE. Key elements of OBE are covered, including the definition of learning goals like Program Educational Objectives (PEOs), Program Outcomes (POs), and Course Outcomes (COs). It also discusses assessment methods—both direct and indirect—and the role of continuous quality improvement through feedback. OBE helps institutions design courses that focus on what students are expected to do with their knowledge, making learning more practical and measurable. Finally, the chapter examines how technology, particularly artificial intelligence (AI), is poised to enhance and automate OBE systems. This preview sets the stage for deeper discussions in the rest of the book.

Keywords Outcome-based education · Student-centered learning · Course outcomes · Assessment methods · Machine learning

1 Introduction

Outcome-Based Education is an educational framework that focuses on **what students should be able to do after completing a course or program**. Instead of concentrating only on what teachers teach, OBE emphasizes what students learn

D. Vetrithangam (✉) · P. Kumar · J. S. Batth
Department of Computer Science & Engineering, University Institute of Engineering, Chandigarh University, Mohali, Punjab, India

D. Vetrithangam et al. (eds.), *Transforming Outcome-Based Education with Machine Learning*, https://doi.org/10.1007/978-981-95-4925-2_1

and how well they can apply their knowledge in real-life situations. OBE is built on **three fundamental principles** that ensure students achieve meaningful learning experiences:

1. **Learning outcomes are clearly defined at the beginning.**
2. **Teaching and assessment methods are designed to help students achieve these outcomes.**
3. **Continuous improvement is made based on student performance and feedback.**

In OBE, before teaching begins, educators define what students should be able to do upon completing a course or program. These are called learning outcomes. A learning outcome is a specific, measurable statement that describes the knowledge, skills, and abilities students will acquire upon completing the course. Outcome-Based Education (OBE) establishes a structured framework by defining learning outcomes at three key levels (Syeed et al., 2022). PEOs specify the broad, long-term goals that the program plans for its graduates to achieve within a few years of graduation, such as career promotions or contributions to society (Tshai et al., 2014). These are a set of essential knowledge, skills, and competencies that a student must possess as a minimum outcome of a degree course, to meet corporate expectations and the industry demand. At a finer level, COs specify the defined learning outcomes that students should have accomplished upon successful completion of an individual course, thereby supporting the larger goals of the program. Such an understanding of the hierarchical structure ensures that there is no overlap of curricula, thus each part of the curriculum contributes meaningfully to both student development and professional preparedness. This work is intended for educators, academic researchers, and policymakers seeking to understand how Machine Learning (ML) can be integrated into Outcome-Based Education (OBE). By combining theoretical foundations, technical approaches, and policy-level perspectives, the manuscript provides insights that support both academic study and practical implementation.

1.1 How OBE Aligns with Global Educational Frameworks and Accreditation Bodies?

OBE is a student-centered approach that focuses on measurable learning outcomes rather than traditional time-based learning models. OBE ensures that students are trained in specific competencies, skills, and knowledge, enabling them to graduate and be prepared to enter the workforce and compete globally. OBE principles have been adopted by many international accreditation bodies and educational frameworks to ensure high-quality education standards, recognition of the degrees abroad, and cross-border mobility of such graduates. Accreditation agencies assess universities and their capacity to fulfill well-defined educational aims by ensuring that graduates are capable of satisfying defined educational goals. OBE has been incorporated into the evaluation criteria of several accreditation bodies worldwide, as detailed in the following subsections.

1.1.1 Washington Accord (For Engineering and Technology Programs: Global)

The Washington Accord is an agreement between accreditation organizations in different countries that assigns specific educational standards to the engineering programs of these nations (Cubos, 2025; Ho et al., 2024). In 1989, the year it was formed, it was also decided that accredited engineering degrees would be mutually recognized, and therefore, professional practice and further studies would be unfettered. Graduates still need to demonstrate specific Program Outcomes (POs) that, in addition to technical proficiency, encompass professional domain skills such as problem-solving, ethical practice, teamwork, and effective communication. Such skills are developed through OBE's emphasis on real-world projects, case studies, and other industry experiences. Graduates also assume ethical responsibilities to themselves, the community, and the nation in scenarios such as construction and power plant projects, with a special emphasis on safety and sustainability. Problem-solving, effective communication, and teamwork are essential for engineers (mechanical, electrical, software, and design) who work on interdisciplinary projects, such as autonomous vehicles.

1.1.2 ABET (Accreditation Board for Engineering and Technology: USA)

The Accreditation Board for Engineering and Technology (ABET) is a globally recognized agency for accrediting Science, Technology, Engineering, and Mathematics (STEM) programs. It ensures that educational programs demonstrate the results of the model used to prepare students for professional success. For instance, a university in the USA may seek ABET accreditation for its Computer Science program (Mane, 2024). It will need to define and assess the key knowledge outcomes the program aims to achieve, including the ability to design, build, and analyze complex computing solutions, as well as the capacity to work in multidisciplinary teams to solve real-world engineering problems (Shuman et al., 2005). Programs that become accredited will require the use of a continuous improvement process to achieve this, including the measurement, analysis, and refinement of their student outcomes (SOs) as part of regular practice to ensure high-quality education.

1.1.3 NBA (National Board of Accreditation: India)

India has a single body, which accredits engineering, management, and applied science programs, known as the National Board of Accreditation (Goldaper, 1987). In line with the Washington Accord, the structures are designed to ensure that Indian engineering graduates achieve international standards. To achieve NBA accreditation, students who complete B. Technical Electrical Engineering requires specialized learning and evaluation of skills to prepare individuals for designing electrical systems, while also developing essential soft skills such as teamwork and maintaining awareness of ethical and environmental considerations. The process of the NBA quality education requirement requires institutions to establish POs and COs and perform direct and indirect outcome evaluation for student achievement verification.

1.1.4 NAAC (National Assessment and Accreditation Council: India)

NAAC is India's national accreditation body that assesses higher education institutions on **teaching quality, research capability, and community engagement**. It requires institutions to adopt **OBE principles**—primarily demonstrated through student outcome tracking, employability enhancement, and curriculum improvement cycles. Unlike the Washington Accord's focus on global engineering mobility, NAAC emphasizes **institution-level performance** across academic and administrative functions, demanding evidence of continuous improvement (Kumar et al., 2023). Increasingly, NAAC accreditation is supported by **AI- and analytics-based tools that personalize student learning paths, forecast performance, identify areas for improvement, and provide faculty with feedback to enhance their** pedagogy. Through digital platforms, institutions enhance curriculum delivery and ensure alignment with national education goals and employability requirements. Table 1 presents a comparative view of accreditation bodies and their alignment with OBE principles.

1.1.5 Comparative Perspectives on Global OBE Frameworks

Globally, OBE frameworks exhibit both convergence and regional variation. In Europe, the European Qualifications Framework (EQF) and the Bologna Process standardize levels, credits, and quality assurance (QA), promoting mobility and recognition, but face uneven national implementation. Africa combines continental initiatives, such as Tuning Africa, with strong national qualifications frameworks, emphasizing employability and the recognition of prior learning (RPL), although constrained by resource and quality assurance capacity gaps. Association of

Table 1 Accreditation frameworks and relevance to OBE

Accreditation Body	Key Focus/Criteria	Relevance to OBE
NBA (India)	Program outcomes, continuous improvement	Aligns directly with OBE outcome-based criteria
ABET (USA)	Student outcomes, assessment methods	Emphasizes measurable competencies, supporting OBE
Washington Accord	Graduate attributes, global mobility	Promotes international recognition of OBE standards
NAAC (India)	Curriculum design, teaching-learning, and evaluation	Encourages outcome-based curriculum delivery

Southeast Asian Nations (ASEAN) advances through the ASEAN Qualifications Reference Framework (AQRF) and ASEAN University Network–Quality Assurance (AUN-QA), with countries such as Malaysia and the Philippines mandating OBE, demonstrating precise alignment of PEOs, program learning outcomes (PLOs), and course learning outcomes (CLOs), while struggling with harmonization across diverse systems. Latin America, guided by Tuning and Accreditation System for Undergraduate University Degrees in the Southern Common Market (MERCOSUR) and Associated States (MERCOSUR's ARCU-SUR), *stresses* competencies and evidence-based assessment, yet faces variability in quality assurance maturity and resource limitations. Collectively, these regions demonstrate a shared commitment to mobility, comparability, and learner-centered outcomes, while differing in terms of implementation depth, quality assurance mechanisms, and capacity for large-scale assessment.

2 Historical Background and Genesis of OBE

OBE represents an educational framework that defines specific learning achievements through identical assessment methods. Teachers at the beginning of the twentieth century abandoned content-driven instruction in favor of student-centered education. The U.S., along with Canada and Australia, demonstrated a deep interest in this educational concept throughout the 1980s and 1990s, as educational reforms increasingly demanded quantifiable student accomplishments. OBE serves as an assurance method to produce students who acquire the necessary knowledge along with skills and competencies that become critical in real-world activities (Hejazi, 2011). Traditional education relies on content transmission, whereas OBE defines learning objectives and founds its assessment approach on student performance outcomes. The widespread use of OBE across engineering, medicine, and professional studies has occurred recently due to its positive impact on student employability and continuous learning. Modern education systems now exceed traditional content-based teaching approaches. Outcome-Based Education (OBE) and Competency-Based Education (CBE). A core objective behind this transformation seeks to deliver learners access to practical abilities, critical thinking skills, and practical competencies for real-world applications.

2.1 Traditional Education Model

For many years, traditional education has been the main method used to teach children. As shown in Figure 1, this system was intended for the delivery of structured content and standardized assessments. Although this model has produced several successful professionals, it has a few drawbacks in preparing students to address real-world problems (Pakdaman et al., 2019). Traditional education has the following key characteristics:

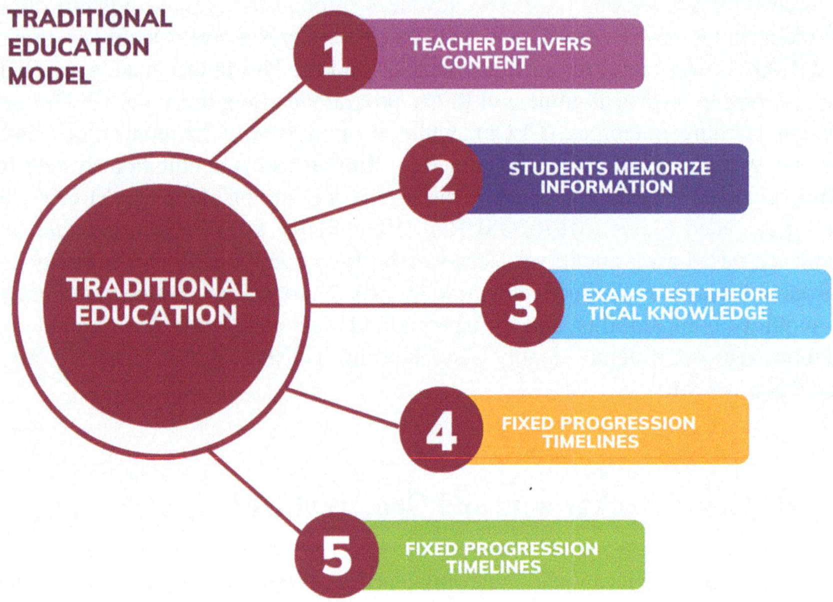

Fig. 1 Structure of the traditional education approach

1. Content Delivery Through Lectures

   Students in traditional educational environments typically receive all their classroom knowledge through teacher-led lectures, as the teacher serves as the primary source of content. The teaching process delivers information to learners who must take notes while preparing for standardized assessments. Lectures serve as a strong method for fast information transfer, yet they lack interaction with students who might struggle to develop logical thinking skills.

2. Fixed Learning Timelines

   Typical education follows a strict schedule, where all students progress through the program together without considering differences in their educational needs. Unlike this approach, which does not account for how students understand concepts—some grasp them quickly, while others take more time, and the pace is too slow for them. The system prioritizes completing courses over actually mastering the skills and concepts that are meant to be taught.

3. Standardized Assessments and Memorization

   In the traditional education system, assessment plays a crucial role, typically conducted through standardized tests, written examinations, and quizzes. Typically, these assessments rely too heavily on rote memorization rather than application or deep understanding. Learning is often focused on recalling facts and reproducing information, but it does not necessarily reflect students' ability to apply the skill or solve problems. Therefore, this approach restricts the possibilities of experiential learning and creative thinking.

4. Teacher-Centered Learning with Limited Student Engagement

   The traditional teacher-directed school environment controls both content choices and instructional methodology, as well as the instructional timeline. Students receive few active learning opportunities that allow them to discuss with peers or work collaboratively, as well as to learn through practical experience. The improper execution of this approach can lead students to focus more on exam success rather than actual learning.

However, does this approach effectively prepare students for real-world challenges? Critics argue that traditional education often fails to develop problem-solving skills and adaptability, necessitating a more structured and outcome-driven approach to education.

2.2 *Transition to Outcome-Based Education (OBE)*

The educational shift from traditional teaching techniques toward learning outcome-based education is known as Transition to Outcome-Based Education (OBE). The teaching methods, along with assessments and the curriculum within OBE, are purposefully designed to help students acquire essential knowledge and skills before graduation. Through OBE, learning institutions measure student progress and performance toward well-defined results that prepare graduates for real-world activity. The orientation to outcome-based education requires educational institutions to restructure course designs, faculty training methods, and assessment mechanisms that match the target educational outcomes (Zamir et al., 2022). Figure 2 illustrates the transition from traditional education to outcome-based education.

2.2.1 Clearly Defined Learning Outcomes

OBE means moving from the 'you learn this by this' objective to 'these are the desired learning outcomes.' Outcomes describe what students should know, be able to do, or demonstrate when they finish a course or program. Institutions establish these outcomes in response to industry demands, real-world applications, and the development of higher-order thinking skills.

For example, instead of just covering a chapter on programming in a computer science course, an OBE approach would set an outcome such as students will be able to design and implement an optimized sorting algorithm. It helps the students get practical skills as opposed to theoretical knowledge.

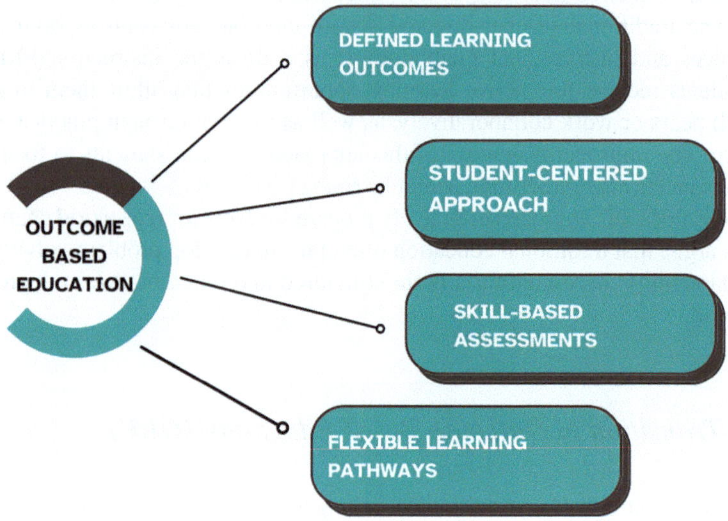

Fig. 2 Transition from traditional to outcome-based education

2.2.2 Student-Centered Learning

Traditional models that control the curriculum through the teacher are replaced by OBE, which puts the student at the center of focus, allowing them to participate in their own learning and become more involved. The learning activities are designed to foster critical thinking, problem-solving skills, and hands-on experience. For example, in an OBE-based engineering course, the students may be engaged in creating a real-world project on a smart irrigation system rather than learning from a lecture on irrigation techniques.

2.2.3 Continuous Assessment Based on Skill Application

In Outcome-Based Education (OBE), both formative and summative assessments are used to track how well students can apply what they have learned. Unlike traditional systems, OBE does not rely solely on exams to measure success

- Project-based assessments
- Case studies and problem-solving exercises
- Practical demonstrations
- Industry-relevant assignments

For example, a business management course could examine students by having them design and present a business strategy for a new business, rather than simply answering theoretical business strategy questions on an exam.

2.2.4 Flexibility in Learning Pathways

Students can progress through OBE at their own pace, as there are various pathways for achieving the set learning outcomes. It is highly flexible, being fit for individualized learning experiences and learning styles.

Example: OBE-driven language learning program: In other words, students have freedom in choosing different methods of learning like online interactive modules, group discussions, or immersion programs, yet lead to the same outcome—language proficiency.

2.3 Evolution into Competency-Based Education (CBE)

Competency-Based Education (CBE) enhances OBE by requiring students to demonstrate mastery of specific competencies before program progression. OBE principles apply in this context, yet the system focuses primarily on individual advancement while dealing with genuine world phenomena. CBE offers independent learning options, enabling each student to adapt their pace for challenging content, allowing them to spend adequate time mastering these areas before moving beyond them. The flexible method promotes better learning outcomes for students with diverse needs. Under CBE, students must demonstrate an understanding of the material before moving forward, as the system does not rely on scheduled classroom attendance. (Irvine & Kevan, 2017). That way, they have a chance to fully develop the skills before upgrading. For instance, in the medical program, students will be required to demonstrate their ability to diagnose diseases using simulated patient cases before progressing to more complex clinical training. On the other hand, CBE also emphasizes solving real-world problems by instigating the application of knowledge gained by students in practical situations. Everything is based on hands-on projects and real-world challenges rather than memorization. For example, in a cybersecurity course, students might be evaluated by their ability to identify and respond to simulated cyberattacks, rather than taking a traditional textbook-based exam. As shown in Figure 3, Competency-Based Education (CBE) focuses on skill-based learning, mastery before progression, industry-aligned outcomes, and real-world application. The model ensures learners achieve defined competencies at their own pace, aligning education closely with practical and industry needs.

This approach helps ensure that not only do students understand the concepts, but also that they can apply those concepts practically in their daily lives. Whereas traditional education does not *stress* the goals that learning should achieve, OBE does. However, what are the implications of this on student engagement and their long-term retention of learning? According to research, OBE helps graduates develop a deeper understanding and application of knowledge, making them more ready for the industry.

However, what is the difference between OBE and CBE in practice? Although both emphasize outcomes, CBE guarantees that students cannot progress to the next

Fig. 3 Competency-based learning approach

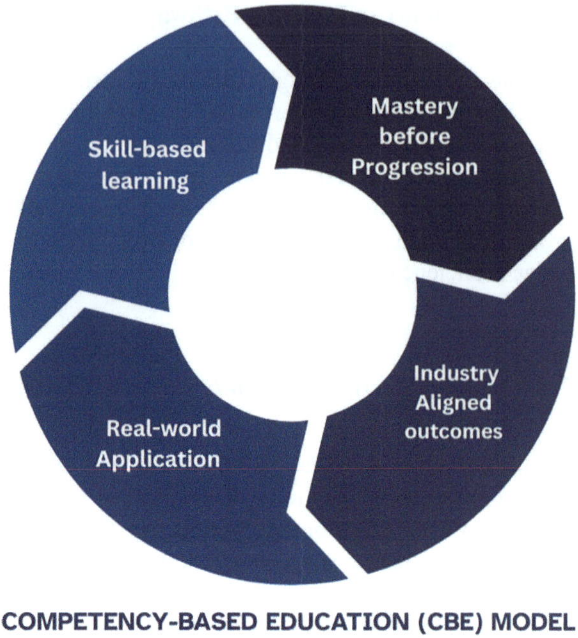

level unless they demonstrate. In this case, they show mastery of a competency, and they are very rigorous and application-driven. For example, in a medical education program, no student is allowed to progress to a higher level without proving their competency in diagnosing diseases by solving case studies, rather than by completing a fixed number of classroom hours. This ensures that the medical healthcare professional is thoroughly prepared.

2.4 Trace the Origins of OBE from Traditional Education Models to Competency-Based Education

Traditional models of education gave way to a more recent development known as Outcome-Based Education (OBE). The previous education was teacher-centered, in which students sat in an auditorium and took part in examinations, memorized textbooks, and learned through lectures (Voorhees, 2001). The behaviorist learning model in the 20th century emphasized structured lessons and reinforcement, but also placed a strong emphasis on rote learning. When you go back to the 1950s, and you are talking about Bloom's Taxonomy, everyone was encouraged to apply those different levels of learning, thinking, problem-solving, and developing skills. Competency-Based Education (CBE) emerged as a term that enables students to demonstrate real-life abilities instead of completing all coursework in classes. OBE suggested that learning should maintain clear educational objectives because this method teaches the practical skills necessary for students to succeed. OBE stands in contrast with traditional education because it enables students to study at their own pace as long as they achieve and demonstrate the related learning outcomes. Currently, OBE is being applied in schools, universities, and professional training, where learners face real challenges and work on their solutions.

2.5 Global Reforms and Pioneers in Outcome-Based Education

Several countries have adopted Outcome-Based Education (OBE) through various reforms to improve student learning and skill development. Performance-based education in the United States began in the 1980s, assessing students not by the time they spend in school, but by the skills they have learned. In portfolios, students demonstrated learning through projects instead of standard exams, which schools used to prepare them for the real world better. In 1990, Australia implemented a Competency-Based Training (CBT) framework within vocational and higher education to enhance the practical orientation of learning (Smith, 1999). The assessment method examined nursing students through both theoretical knowledge evaluation and practical emergency handling skills to ensure they met employment standards. South Africa adopted Curriculum 2005 in 1997 as an OBE-based reform to establish learning opportunities for everybody. The educational method required students to analyze historical events rather than memorize facts to build their critical thinking ability. The Bologna Process of the European Union introduced its credit system in 1999, which required students to demonstrate their subject readiness, rather than relying solely on classroom lectures. The reform bridged various nations and students by allowing them to acquire credentials. In 2020, India implemented the National Education Policy (NEP) 2020, which encourages project and problem-based learning as opposed to rote memorization. In the NEP, students as young as 12 were engaged in projects that addressed sustainable solutions to problems in their neighborhoods. These reforms in other countries have made education more practical and skills-based, to prepare students to tackle real-world challenges. The design of Outcome-Based Education (OBE) has been influenced by numerous educators and researchers who have pioneered various learning paradigms that focus on outcomes, skills, and real-world contexts. These new ideas were introduced by William Spady, Benjamin Bloom, and John Dewey, and to a certain extent, drastically transformed the traditional education system by focusing more on the learners and the skills to be attained.

2.5.1 William Spady: The "Father of OBE"

William Spady is an American education reformer most widely known as the 'Father of Outcome-Based Education'(Tucker, 1998). Additionally, he introduced modern OBE frameworks as a departure from the traditional conception of teaching, with a focus on student learning outcomes. While most systems require students to progress through completing a specified period of time (such as grade levels), Spady advocated that students should only advance after an acceptable level of skill and knowledge has been acquired.

For instance, in a traditional school, a student who does not understand key math concepts would still be promoted to the next grade level. In an OBE system, however, the student would have to continue learning the material until they demonstrate full competence before they could move on. With this method, all students are assured of acquiring the relevant skills, rather than merely attaining a passing mark, as they are not just memorizing content.

As a result of Spady's work, four key principles of OBE were established.

1. Defines the clear learning outcomes that students are required to achieve.
2. Designing Backward—Create lesson, unit, and course outlines from the intended outcomes.
3. High Expectations—They believe that all students can be successful with proper support.
4. Expanded Opportunities—An opportunity for advanced learning for students who move fast.

He had his ideas implemented widely in the United States, South Africa, Canada, and Australia, and these influenced major education reforms.

2.5.2 Benjamin Bloom: Creator of Bloom's Taxonomy

Cognitive levels of learning are structured in a hierarchy of levels, as outlined in Bloom's Taxonomy, developed by educational psychologist Benjamin Bloom (Bloom, 1956). Before that assignment, he had been using rote memorization techniques in which students were assessed on their ability to recall facts. The learning model of Bloom's Taxonomy features hierarchical steps that promote the development of advanced cognitive skills.

Bloom's Taxonomy has six levels as follows:

1) Remembering—Recalling facts, basic concepts, or previously learned information.
2) Understanding—Explaining ideas, concepts, or interpreting meaning.
3) Applying—Using acquired knowledge in new and practical situations.
4) Analyzing—Breaking down information into components to understand relationships and structures.
5) Evaluating—Justifying decisions, critiquing, or making informed judgments.
6) Creating—Producing new work, ideas, or original products.

Take a science class as an example; under the traditional approach, students may be required to memorize the periodic table (Remembering). However, with an OBE-based approach, students would be able to apply their knowledge through experiments (applying) and analyze chemical reactions (analyzing). In doing so, Bloom's work had a profound influence on OBE by ensuring that education extended beyond mere memorization to thinking critically, solving problems, and fostering creativity, making learning more meaningful and applicable in real-world situations.

2.5.3 John Dewey: Philosopher of Experiential Learning

John Dewey defined experiential learning as an active learning process that contrasts with passive information absorption, as outlined in his work as a philosopher and educational reformer (Miettinen, 2000). His academic approach relied on hands-on learning to create an educational system that was student-centered. For example, instead of studying farming through textbooks, Dewey's teaching would allow students to visit a farm, plant crops, and even watch the crops grow. Likewise, a history class could involve having students reenact historical events or discuss different perspectives rather than memorizing dates and facts. OBE followed in line with Dewey's philosophy, which promoted a hands-on, project-based approach to learning that instills students with the skills to face real-world challenges. His concepts currently guide modern efforts, including educational structures, vocational training, and STEM programs, which teach skills applicable to students' career prospects and daily activities.

3 Evolution and Global Adoption of OBE in Higher Education

Outcome-Based Education (OBE) is a contemporary educational approach that focuses on what students should be able to achieve upon completing a course or program. While traditional learning typically aims to transfer content and is measured by the time spent in class (input-based learning), OBE ensures that students acquire specific skills, knowledge, and competencies (outcome-based learning).

3.1 How OBE is Adopted Across Higher Education Worldwide?

OBE is a student-centered learning approach that emphasizes learning outcomes rather than merely completing the syllabus. Universities and colleges worldwide have adopted Outcome-Based Education (OBE) to improve educational quality, align learning with industry needs, and boost students' competitiveness in the global job market. Universities integrate Outcome-Based Education (OBE) by designing curricula with clear learning outcomes for each course, ensuring students acquire relevant knowledge and skills. In recent years, traditional teaching methods have been replaced with active learning, problem-solving, case studies, and hands-on projects, rather than relying solely on memorization. Traditional exams are replaced with assessments that incorporate rubrics, projects, presentations, and skill demonstrations to determine whether students know and can apply the material. Engineering programs worldwide have adopted Outcome-Based Education (OBE) to ensure that graduates meet international accreditation standards. Engineering colleges in the United States are mandated by organizations like ABET and India by NBA to follow an outcome-based framework; i.e., to develop required skills and competencies for the industry.

3.2 Milestones in the Implementation of OBE

3.2.1 Shift from Input-Based to Outcome-Based Assessment

Previously, education was focused on the duration of classes, the materials students were required to read, and instruction that was teacher-led. However, Outcome-Based Education (OBE) shifts the focus to what students actually learn and can apply in real life. Students learn practical skills and competencies, rather than just memorizing theories. For example, it is known that in a traditional input-based system, any student at the university was required to listen to 100 h or more of physics lectures before graduating, regardless of whether they grasped the concepts or not. On the other hand, students who practice OBE are tested on their ability to apply physics principles to solve real-life problems, such as designing a stable bridge or an energy-efficient home. It ensures that the learned material is relevant and can be applied directly to real-life situations.

3.2.2 Global Milestones in OBE Implementation

Outcome-Based Education (OBE) has evolved over the years, shaping how students learn and are assessed worldwide. Table 2 shows the milestones in the implementation of OBE. The Washington Accord, as it is known, establishes global standards

Table 2 Milestones in the implementation of OBE

Year/ Period	Milestone/Event
1980s	The **Washington Accord** was established, setting international standards for OBE in engineering education.
1990s	Australia, Canada, and the USA introduced OBE in schools and universities.
2000s	**The Bologna Process (Europe) unified higher education standards across Europe by adopting the OBE approach.**
2010s	Many Asian countries (India, China, Malaysia, Singapore) adopted OBE for university accreditation.
2020s	AI-driven learning analytics help measure student outcomes in OBE frameworks worldwide.

for engineering education and engineering qualifications worldwide, and was established in the 1980s (Oladiran et al., 2013). OBE was introduced in Australian schools and universities by the late 1970s (Donnelly, 2007), in Canadian schools by the late 1980s, and in schools and colleges in the USA by the 1990s (Davies & Hammack, 2005). Starting from the 2000s, the Bologna Process implemented the principles of OBE in European higher education, standardizing them (Gaebel et al., 2018). In the 2010s, many Asian countries, especially India and China, adopted OBE for university accreditation (Qadir et al., 2020). Currently, AI-powered learning analytics are improving a school or institution's ability to measure students' outcomes more effectively, thereby making the impact of OBE more data-driven and effective.

4 Core Principles and Components of OBE

Outcome-Based Education (OBE) focuses on what students should achieve by the end of their learning journey. OBE considers well-defined learning outcomes, rather than time spent in classrooms, to be a sign of success. These outcomes are categorized at different levels, including PEOs, POs, and COs. With this approach, students acquire the necessary practical skills, knowledge, and competencies that align with the needs of industry and the real world.

4.1 Program Educational Objectives, Program Outcomes, Course Outcomes.

PEOs are the goals that program graduates are expected to achieve 3–5 years after program completion. This set of objectives is designed to align with industry needs, higher education, and serve the greater good of society. For instance, in a Computer Science Engineering (CSE) program, the PEOs could, for example, state that graduates will work in top IT companies for software development, pursue higher education or research in advanced computing, or become entrepreneurs and set up a technology-based business. PEOs represent professional targets that students need to accomplish during the period after finishing their studies. POs are the skills and knowledge that students are expected to achieve upon graduation. The focus of these outcomes is on technical expertise, problem-solving skills, teamwork, communication, and the application of knowledge in a practical manner. For example, in a Mechanical Engineering program, students should be able to apply mathematics and physics to solve engineering problems, design machine components using modern design software, and collaborate in a team to complete mechanical engineering projects. POs focus on the essential skills and knowledge that students are expected to have acquired by the time they complete their program. Specific learning goals for each subject in a program are referred to as COs. They tell you what students should understand and be able to do upon completion of a course.

4.1.1 Formulation of Course Outcomes for Specific Courses

Formulating COs is a critical part of designing any curriculum within the Outcome-Based Education (OBE) framework. COs are clear, concise statements that describe what a student is expected to be able to do after completing a course. As shown in Figure 4, the process begins with a solid understanding of the broader educational context, including the university's several challenges and criticisms, as well as the POs and Program Specific Outcomes (PSOs) defined for each academic program. These foundational elements guide the course's vision and direction. Additionally, factors such as current industry needs, societal challenges, and emerging trends should also be considered. Increasingly, educators are encouraged to incorporate global priorities, such as the United Nations Sustainable Development Goals (SDGs), particularly when course content has a societal or environmental impact.

Once the broader goals are established, the next step is to define the course objectives—general statements that outline what the course aims to teach. These are not yet measurable, but they set the direction for framing more specific and assessable outcomes. The actual CO formulation begins by identifying the appropriate cognitive level for each learning expectation using Bloom's Taxonomy (BT). Bloom's Taxonomy provides a structured hierarchy of thinking skills, ranging from basic knowledge recall to complex tasks such as evaluation and creation. Based on the desired level of learning, action verbs are selected—these are crucial because they ensure that each CO is measurable and observable.

For instance, verbs like "explain," "analyze," "design," or "implement" are far more useful than vague words like "understand" or "learn." A well-framed CO typically follows a structure that includes what the student will do, how they will do it, and for what purpose. Table 3 presents some examples CO statements for the Object-Oriented Programming using Java course. Typically, around 3 to 6 COs are created, depending on the level of detail or breadth of the course content. Framing well-structured and slightly generic COs is considered a good practice, as it allows

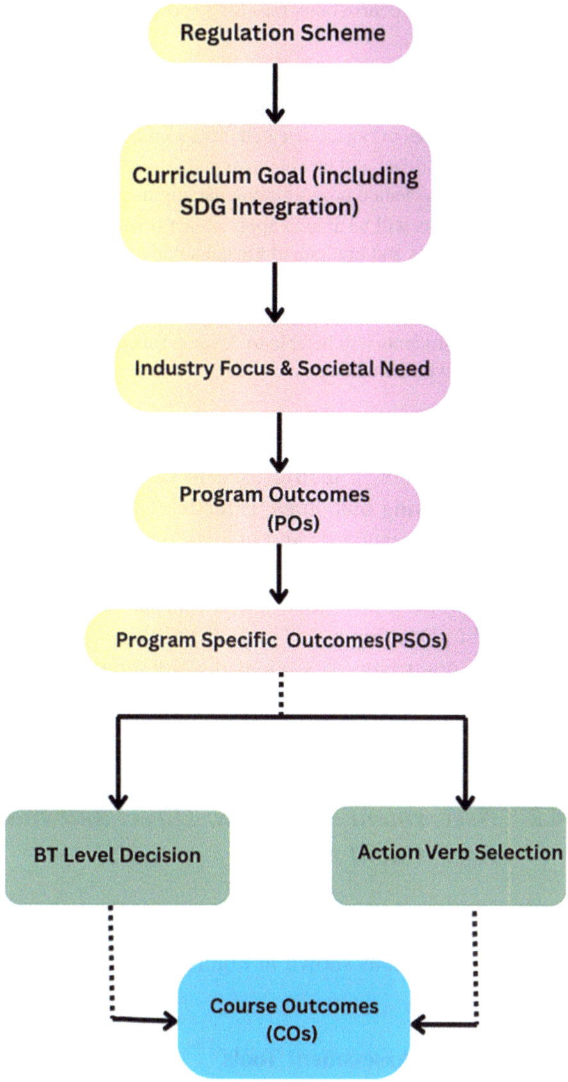

Fig. 4 Course outcome formulation framework

for easier adaptation across related subjects. These outcomes should be attainable within the course duration and must align directly with one or more of the POs or PSOs. This alignment is often documented using a CO-PO mapping matrix, where each CO is linked to relevant POs and evaluated for its contribution on a scale of low, medium, or high. In addition, COs may also be mapped to specific SDGs if the course contributes to global sustainability or ethical responsibility—for instance, a course on renewable energy could support SDG 7 (Affordable and Clean Energy) or SDG 13 (Climate Action).

To ensure that COs are not only well-framed but also actionable, appropriate assessment methods must be established. Each outcome should be assessed through

Table 3 Example course outcomes for a specific course

CO No	Course Outcome Statement
CO1	The students will be able to recall and understand the basic concepts, terminology, and principles of object-oriented programming.
CO2	The students will be able to explain the features, functionalities, and key concepts of object-oriented programming paradigms using Java.
CO3	Students will be able to apply object-oriented programming concepts to solve real-world problems and implement logical solutions using Java.
CO4	The students will be able to evaluate and apply advanced programming concepts to demonstrate their functionalities and practical applications
CO5	The students will be able to develop functional and interactive applications to address real-world challenges.

quizzes, assignments, lab work, projects, or exams. This facilitates the calculation of CO attainment, a metric used to determine whether students are meeting the expected learning outcomes. Effective COs are specific, measurable, aligned with both academic standards and industry needs, and supported by a robust assessment strategy. By following a systematic approach—grounded in Bloom's Taxonomy, guided by program goals, and optionally enriched through alignment with Sustainable Development Goals (SDGs)—educators can ensure their courses significantly contribute to student development, institutional excellence, and global relevance.

4.2 Assessment Strategies: Direct and Indirect Assessment Methods

In Outcome-Based Education, CO attainment is evaluated using direct and indirect assessment tools, as shown in Figure 5.

4.2.1 Direct Assessment Tools

Student performance determines how direct assessment tools reveal what students know and are capable of doing. The direct assessment method comprises Monthly tests, Tutorials, Assignments, Quiz, Mini Project, Seminar and Technical Presentation as well as other assessment components that link to specific COs in each course. The assessment tools applied to course outcomes can also be directly applied to program outcomes and specific program outcome assessments.

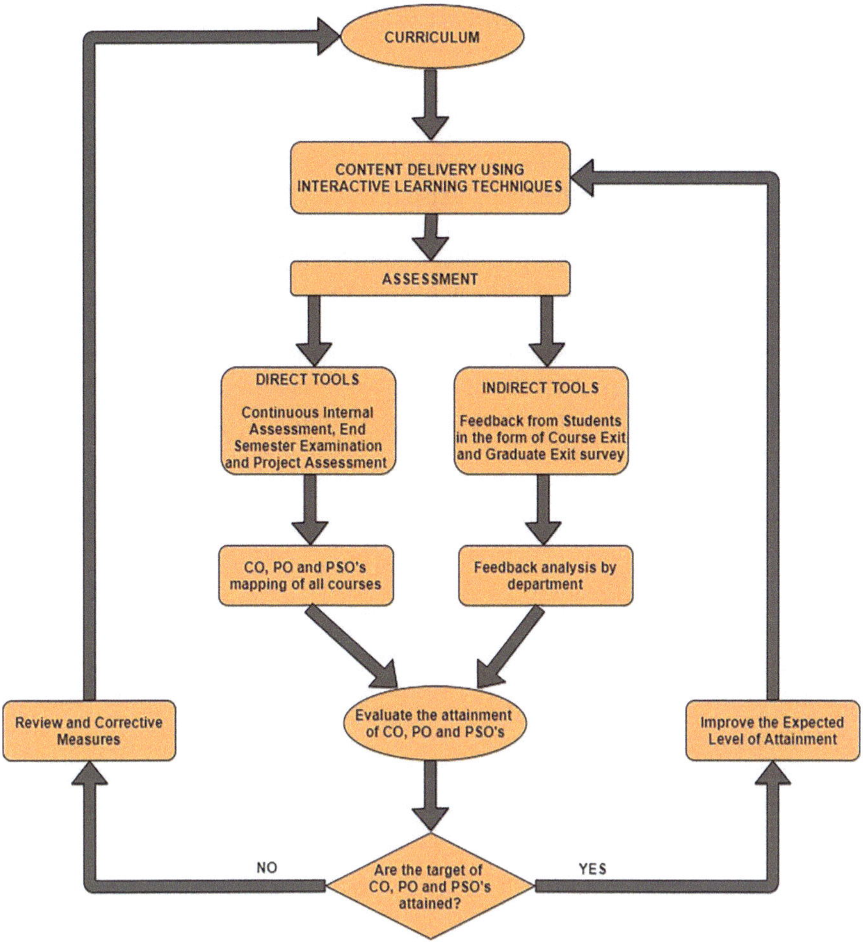

Fig. 5 Process flow for assessing curriculum compliance with POs and PSOs

4.2.2 Indirect Assessment Tools

Attainment of PO and PSO is also measured through indirect assessment tools. Indirect assessment encompasses feedback obtained from students through graduate exit surveys, as well as feedback from parents, employers, and faculty.

4.3 Continuous Quality Improvement (CQI) Through Feedback Mechanisms

The ongoing systematic process of CQI works to enhance both educational practices and student learning outcomes and institutional performance. The method consists of routine assessments that follow existing process examinations and subsequent identification of potential enhancement points, leading to better outcomes during successive time periods. It aims to collect feedback, pinpoint problems, and enhance its effectiveness and efficiency, ultimately improving customer satisfaction (Roffe, 1998). Therefore, feedback mechanisms are the means by which areas that need improvement are identified. Surveys, performance reviews, customer feedback, sensor data, and automated reports can be used as these mechanisms. This information, like any other, is used by organizations to make informed decisions and continuously improve their processes.

For example, if many students fail a specific subject, the institution takes measures to change its teaching methods and assessment strategies to improve the results.

5 Challenges and Criticism of OBE

OBE focuses on student learning outcomes rather than just the teaching process. It attempts to ensure that students learn to apply knowledge, skills, and abilities in real-world applications. However, there are several challenges and criticisms associated with implementing OBE.

5.1 Implementation Difficulties in Different Education Systems

Implementing OBE is challenging because it necessitates a comprehensive overhaul of the educational structure. Each country and each institution have its own rules, different resources and different ways of teaching. This poses difficulties to implement OBE in the same way across. There are certain institutions which may possess the requisite faculty trained in OBE or requisite technology but may not be able to adopt OBE because of the reason of unavailability of resources or rigid traditional systems. Therefore, the application of OBE has been a complex problem across different education systems. The problem in education is to provide a well-trained teacher, who can keep their students from disengaging, and to grade courses fairly and consistently. Some teachers may not be familiar with modern teaching methods, such as OBE or interactive learning, which can impact how effectively students learn. Teachers can participate in training programs and workshops to refine their skills. Another issue is student engagement. Sometimes, traditional lectures can be, well, boring, and it is difficult for students to learn from them. By utilizing group discussions, hands-on activities, and real-world projects, learning can become both more engaging and effective. A reason for this concern is the grading system, as different teachers may evaluate students in varying ways. And schools should grade fairly, with uniform standards and have uniformed assessment procedures.

One topic within educational discussion is that of traditional teaching or OBE. Conventional education, on the other hand, is based on classroom lectures and exams, where students learn by rote to pass exams. This kind of approach ensures that students have a solid knowledge foundation, but they may not be prepared for real-world scenario applications. On the contrary, OBE focuses on what students are expected to do after completing a course. In particular, it motivates the development of technical knowledge through projects and skill-based assessments. OBE ensures that students learn practical skills but teachers become overburdened with the elaborate plans and methods of teaching. It's best to combine both methods in a good educational system, utilizing the best aspects of each to provide students with a well-rounded experience.

6 The Role of Technology in Advancing OBE

The digital transformation of OBE is done using Learning Management Systems (LMS). An online platform aiding schools and universities to manage courses, monitor student progress and hand out digital content is termed as a LMS. It plays a key rolc in OBE by cnsuring students achieve learning outcomes through structured content, assessments, and feedback. For instance, a university can choose to integrate Moodle as its LMS to manage educational courses. The lecture videos can be uploaded to the site by professors, along with assignments and quizzes, for their monitoring of a student's understanding. This allows students to submit assignments online and receive instant feedback to improve their performance. Additionally, LMS analytics enable educators to identify students who are struggling in specific areas and allow programmers to create customized solutions to promote learning. Digital assessments and structured learning paths, along with real-time progress tracking offered by an LMS platform, help students meet their learning goals.

6.1 Contribution of Data Analytics, AI, and Machine Learning to OBE Implementation

Data analytics plays a crucial role in OBE by collecting and analyzing student performance data to identify patterns and trends. It helps institutions improve their teaching methods by leveraging insights from data. For instance, if a university questions a quiz score on an exam and then evaluates students in a particular course, such as Data Structures, and finds that 60% of the students are struggling with Dynamic Programming, the professor can take necessary steps, such as conducting extra tutorial sessions to help students better understand the material. Similarly, AI and ML also enhance OBE by customizing learning, recognizing the hole and suggesting real—time improvements to it. Coursera and Duolingo use learning algorithms that changes the learning difficulty based on the student performance. If a student has difficulty with a specific topic, AI can automatically suggest additional practice questions or video lectures to help them progress at their own pace. As a result, the content is personalized and tailored to each student's individual needs, and learning takes place adaptively. More so, Predictive analytics can determine which students are "at risk" and can therefore, be supported and intervened with appropriately and in a timely manner. Down the line, it would deploy AI for the purpose of creating fully automated and individualized learning paths, custom virtual reality-centered learning environments, and rapid detection of students' learning difficulties, while also assisting educators in elevating their engagement. AI is making education more intelligent, flexible, and universally accessible.

7 Machine Learning Revolution in Outcome-Based Education

Incorporating ML into OBE is a major shift from traditional assessment models to intelligent, data-driven educational frameworks that optimize student learning outcomes. This change has accelerated since 2020, driven by advancements in artificial intelligence, the growing volume of educational data, and the need for tailored learning solutions that became apparent during the pandemic. OBE is enhanced by having ML in place, which provides automated mechanisms for tracking, analyzing and predicting learning outcomes in record scales and precision. ML-driven OBE differs from traditional OBE models, which operate through periodic assessments, manual tracking, and reporting, in that it provides adaptive learning pathways, monitors students in real-time, and offers feedback systems to both students and educators. This is a shift in focus from retrospective outcome evaluation to outcome optimization.

7.1 Empirical Evidence of ML-Enhanced OBE Performance

Recent empirical studies demonstrate significant improvements when ML technologies are integrated into OBE frameworks. Research in undergraduate education shows that AI-enhanced OBE models substantially outperform traditional approaches, with students achieving higher average scores and demonstrating improved mastery of learning outcomes. The gains are more specific to complex subjects in which an analytical thinking and problem-solving skills are needed. Adaptive learning systems powered by AI have shown substantial improvements in student performance across diverse educational contexts. Studies utilizing advanced AI tools have explained and noticed improvements in course completion rates and increased student engagement, while quantitative analyses have revealed significant improvements in grades, test scores, and overall academic engagement in AI-enhanced learning environments. Predictive analytics capabilities have proven particularly valuable, with implementations demonstrating high accuracy in identifying students at risk of academic difficulties. This early intervention capability represents a fundamental advancement over traditional OBE models that could only assess outcomes retrospectively, enabling proactive support strategies rather than reactive remediation.

7.2 Machine Learning Technologies Transforming OBE Implementation

7.2.1 Predictive Analytics for Outcome Forecasting

ML algorithms are capable of analyzing large datasets containing information about student performance and predicting outcomes with high accuracy. Predictive modeling using Random Forest, Support Vector Machines, and Neural Networks has successfully predicted student performance, identified at-risk students, and provided tailored intervention plans before issues arise that may hinder the students' success (Cao & Mai, 2025). These systems construct student profiles by integrating various data sources, including attendance records, submission logs, quiz scores, forum activity, and time spent on tasks. Sophisticated algorithms can accurately forecast students' failure with certain learning outcomes, providing the opportunity for the correct intervention to be applied before remedial action is required.

7.2.2 Adaptive Learning Pathways and Personalized Content Delivery

ML-powered adaptive learning systems bend the "one-size-fits-all" paradigm of OBE to suit any standard, marking the next quantum leap in the advancement of technology. These systems utilize classification algorithms to divide learners according to their learning behaviors, regression algorithms to forecast outcome variables, and clustering algorithms to consolidate learners with equivalent learning behaviors. Deep learning (DL) models sift through large and complex educational datasets to discern fine-grained patterns in learners' advancement (Hernández-Blanco et al., 2019) and automatically determine the optimal mix of outcome-achievement content, learning throughput, and instructional modalities, adjusting the levels of outcome achievement that each student can realistically attain. Students in adaptive learning programs accelerated by AI exhibit advancement in test scores when compared to other students receiving direct instruction.

7.2.3 Real-Time Assessment and Feedback Automation

Machine Learning (ML) is capable of replacing the manual, time-consuming evaluation of outcome-based education (OBE) with continuous, automated evaluations that provide immediate feedback on learning outcomes. Assessments powered by AI can produce quizzes, grade them, and provide tailored learning recommendations based on the user's activity. Students' written pieces can be thoroughly assessed and given feedback regarding the automated feedback trained by Natural Language Processing (NLP) on the concepts, writing, and outcomes of the course. Processing and analyzing student performance is almost simultaneous. Predictive reporting on student performance can thus be provided almost instantaneously, allowing the student to immediately understand the concepts they need to work on.

7.3 Cutting-Edge ML Frameworks Advancing OBE

7.3.1 Transformer-Based Models for Student Feedback Analysis

The application of DistilBERT models to examine student feedback and measure learning outcomes has yielded positive results (Saxena & Santhanavijayan, 2024). Indeed, models of this kind offer even more sophisticated sentiment and pattern recognition capabilities than traditional sentiment analysis methods, which enhance the learning experience feedback received from students and provide more useful information than traditional surveys on the effectiveness of OBE. The combination of Local Interpretable Model-Agnostic Explanations (LIME) methods with other models and transformer techniques offers student feedback systems for scholarly papers (Dehghani et al., 2025), explaining in detail the important structures and elements of student sentiment on learning outcomes, achievement, and the invaluable feedback received. Such clarity is essential for OBE, which needs to demonstrate, with usable data, that outcomes have been achieved.

7.3.2 Multi-Modal Learning Analytics

The latest ML frameworks blend behavioral analysis with computer vision and natural language processing technologies to generate holistic portraits of individual student's progress. These systems, equipped with advanced AI, simultaneously assess learners on different outcomes by streaming and analyzing various data types, including recordings of student presentations, essays, and other asynchronous digital behaviors. The efficacy of personalized learning is significantly enhanced through the use of DL techniques when multiple data types are utilized, allowing for the assessment of outcomes such as critical thinking, communication, and collaboration with greater precision.

7.3.3 Federated Learning for Privacy-Preserving OBE

Emerging federated learning approaches allow institutions to collaborate on ML model development while maintaining student data privacy. This enables the creation of more robust outcome prediction models by leveraging collective institutional knowledge without compromising individual student information.

7.4 Implementation Challenges and Solutions

7.4.1 Technical Infrastructure Requirements

The integration of ML into OBE would require substantial IT infrastructure including cloud computing resources, data storage systems, and data streaming capabilities. Research suggests that accessibility of knowledge and the information technology (IT) infrastructure are among the key factors that drive successful implementations across institutions in the education sector.

7.4.2 Faculty Development and Change Management

Stakeholder engagement and organizational culture become pillars in the successful implementation of ML OBE. Challenges such as faculty resistance and lack of professional training remain hurdles that require profound change management and continuous professional development. Concerns related to algorithmic bias, data privacy, and the discrimination and inequity of automated systems must be addressed in ML-based OBE systems. The AI tools and systems must strive to eliminate bias, enhance fairness, and provide an explanation and notice for automated decision systems. There are no simple solutions to algorithmic bias, which reflects a lack of awareness of the need for fairness in automated systems. The trust level during the assessment of outcome is highly influenced by the opacity of algorithm behind the model.

7.5 Future Directions: AI-Driven OBE Evolution

ML is expected to achieve full autonomy in learning systems and ensure education-driven predictive analysis systems constantly optimize educational experiences in real-time. These systems are expected to utilize virtual and augmented reality, as well as AI-based tracking systems, which assess learners in complex simulation environments to record tactile and difficult competencies. These learning systems facilitate real-time record-keeping, and Distributed Ledger Technology (DLT) Blockchain is expected to provide transparent and secure verification of learning outcomes. Mastering sophisticated ML will enable real-time global comparisons of learning outcomes to improve educational outcomes.

8 Conclusion and Future Work

Outcome-Based Education has become a transformative blueprint for higher education by shifting the focus from traditional content delivery to student-centered learning and measurable skill achievement. This approach aligns curricula and program goals with global educational standards and accreditation requirements, ensuring institutions uphold both accountability and relevance. The chapter has traced the historical evolution of OBE, outlined its core principles—including explicit learning outcomes and continuous quality improvement—and summarized its worldwide adoption and success in boosting educational effectiveness. Looking ahead, the future of OBE is being shaped by emerging technologies, including artificial intelligence, machine learning, advanced learning management systems, and immersive digital tools. These innovations are designed to automate assessment processes, personalize learning experiences, and deliver real-time feedback, enabling educators to refine their strategies and support students more effectively and continuously. There is a growing emphasis on leveraging analytics and predictive insights to track student progress, improve curriculum design, and enable proactive support for diverse learners. Future research in OBE should continue to explore its integration across interdisciplinary studies and lifelong learning contexts, as well as examine the long-term impact on graduate employability, skill relevance, and institutional performance. By gathering robust, data-driven insights, policymakers and educators can further enhance OBE strategies and sustain continuous improvement in educational systems. Ultimately, the intersection of OBE and advanced technologies will empower institutions to deliver quality, equity, and adaptability for future generations of learners.

References

Bloom, B. S. (1956). *Taxonomy of educational objective.*

Cao, W., & Mai, N. (2025). Predictive analytics for student success: AI-driven early warning systems and intervention strategies for educational risk management. *Educational Research and Human Development, 2*(2), 36–48.

Cubos, J. (2025). *The Washington Accord: An analytical review of its impact on global engineering education.* Available at SSRN 5151405.

Davies, S., & Hammack, F. M. (2005). The channeling of student competition in higher education: A comparison of Canada and the US. *The Journal of Higher Education, 76*(1), 89–106.

Dehghani, Z., Aslansefat, K., Khan, A., & Akram, M. N. (2025). *Explainability of large language models using SMILE: Statistical model-agnostic interpretability with local explanations.* arXiv preprint arXiv:2505.21657.

Donnelly, K. (2007). Australia's adoption of outcomes-based education: A critique. *Issues in Educational Research, 17*(2), 183–206.

Gaebel, M., Zhang, T., Bunescu, L., & Stoeber, H. (2018). *Learning and teaching in the European higher education area.* European University Association asbl.

Goldaper, S. (1987). *NBA: Washington helps the nets take it easy.* The New York Times.

Hejazi, B. M. (2011). *Outcomes-based education (OBE): A transformational perspective on quality and mobility in higher education.* Outcomes-Based Education: A Transformational Perspective, 1–30.

Hernández-Blanco, A., Herrera-Flores, B., Tomás, D., & Navarro-Colorado, B. (2019). A systematic review of deep learning approaches to educational data mining. *Complexity, 2019*(1), 1306039.

Ho, J., Kortian, V., Huda, N., & Lee, A. (2024). Engineering management education: Washington accord accreditation programs. *Engineering Management Journal, 36*(4), 353–365.

Irvine, C. K. S., & Kevan, J. M. (2017). Competency-based education in higher education. In *Handbook of research on competency-based education in university settings* (pp. 1–27). IGI Global.

Kumar, A., Gawande, A., Kristiawan, M., Finthariasari, M., Viona, E., Tank, M., & Gupta, D. (2023). Perception of faculty regarding the proposed new accreditation norms by the National Assessment and accreditation council (NAAC). *International Journal of Multidisciplinary: Applied Business and Education Research, 4*(9), 3180–3187.

Mane, S. D. (2024). *Accreditations of engineering programs in India: A comparative study of ABET EAC and NBA accreditation.*

Miettinen, R. (2000). The concept of experiential learning and John Dewey's theory of reflective thought and action. *International Journal of Lifelong Education, 19*(1), 54–72.

Oladiran, M. T., Pezzotta, G., Uziak, J., & Gizejowski, M. (2013). Aligning an engineering education program to the Washington accord requirements: Example of the University of Botswana. *International Journal of Engineering Education, 29*(6), 1591–1603.

Pakdaman, M., Moghadam, M. N., Dehghan, H. R., Dehghani, A., & Namayandeh, M. (2019). *Evaluation of the cost-effectiveness of virtual and traditional education models in higher education: A systematic review.* Health Technology Assessment in Action.

Qadir, J., Shafi, A., Al-Fuqaha, A., Taha, A. E. M., Yau, K. L. A., Ponciano, J., Hussain, S., Imran, M. A., Muhammad, S. S., Rais, R. N. B., & Rashid, M. (2020, June). *Outcome-based (engineering) education (OBE): International accreditation practices.* In 2020 ASEE Virtual Annual Conference Content Access. ASEE.

Roffe, I. M. (1998). Conceptual problems of continuous quality improvement and innovation in higher education. *Quality Assurance in Education, 6*(2), 74–82.

Saxena, A., & Santhanavijayan, A. (2024, June). Improving educational assessments: Question difficulty classification with DistilBERT and explainable AI. In *2024 15th International Conference on Computing Communication and Networking Technologies (ICCCNT)* (pp. 1–6). IEEE.

Shuman, L. J., Besterfield-Sacre, M., & McGourty, J. (2005). The ABET "professional skills"—Can they be taught? Can they be assessed? *Journal of Engineering Education, 94*(1), 41–55.

Smith, E. (1999). Ten years of competency-based training: The experience of accredited training providers in Australia. *International Journal of Training and Development, 3*(2), 106–117.

Syeed, M. M., Shihavuddin, A. S. M., Uddin, M. F., Hasan, M., & Khan, R. H. (2022). Outcome-based education (OBE): Defining the process and practice for engineering education. *IEEE Access, 10*, 119170–119192.

Tshai, K. Y., Ho, J. H., Yap, E. H., & Ng, H. K. (2014). Outcome-based education–The assessment of programme educational objectives for an engineering undergraduate degree. *Engineering Education, 9*(1), 74–85.

Tucker, J. A. (1998). The ideology of outcome-based education: An interview with William Spady. *Journal of Research on Christian Education, 7*(1), 5–18.

Voorhees, R. A. (2001). Competency-based learning models: A necessary future. *New Directions for Institutional Research, 2001*(110), 5–13.

Zamir, M. Z., Abid, M. I., Fazal, M. R., Qazi, M. A. A. R., & Kamran, M. (2022). Switching to outcome-based education (OBE) system: A paradigm shift in engineering education. *IEEE Transactions on Education, 65*(4), 695–702.s.

The Future of Outcome-Based Education (OBE): Leveraging Machine Learning (ML) for Adaptive Curriculum Design and Real-Time Learning Monitoring

R. Venkatesh, D. M. D. Preethi, and Arulmurugan Ramu

Abstract In this chapter, we discuss how machine learning (ML) can serve as a robust and efficient aid to bring out and enhance outcome-based education (OBE), thereby improving the quality of education in the current era. There is a 15% increase in demand for personalized learning, resulting in a growing need for an adaptive approach. In a traditional education system, addressing learners' needs is typically challenging. For example, research suggests that 40% of students feel they have very little exposure to personalized learning. ML, an edtech company, has contributed to customized learning and adaptive learning through real-time monitoring and fluid learning pathways. ML has been an approach used by educators and policymakers to accommodate the changing needs of individual learners (ML). While the literature is still developing, case studies suggest student performance improved when participating in OBE peer learning with an emphasis on personalized learning utilizing ML. Improvement was reported for some students and groups of learners by as much as a 25% increase in overall performance. Hence, it demonstrates how an adaptive learning experience can yield effective learning outcomes. Of course, there are ethical concerns, especially in terms of data privacy, and equitable access to educational support—estimates indicate that as many as 30% of educational settings face these challenges. Therefore, transformative and equitable education must occur in conjunction with ML—utilizing accessible and responsive systems—but must also be systematic and responsible in the use of data and

R. Venkatesh (✉)
Department of Physics, PSNA College of Engineering and Technology (Autonomous), Dindigul, Tamil Nadu, India

D. M. D. Preethi
Department of Computer Science and Engineering, PSNA College of Engineering and Technology (Autonomous), Dindigul, Tamil Nadu, India

A. Ramu
Department of Computational Science and Software Engineering, K. Zhubanov University (Heriot-Watt), Aktobe Campus, Aktobe, Kazakhstan
e-mail: a.ramu@hw.ac.uk

D. Vetrithangam et al. (eds.), *Transforming Outcome-Based Education with Machine Learning*, https://doi.org/10.1007/978-981-95-4925-2_2

infrastructure. It will need to continue to be the research that ML is efficacious and equitable in education and beyond.

Keywords OBE · ML · Adaptive curriculum design · Educational technology · Data-driven pedagogy · Student performance prediction · Curriculum innovation

1 Introduction: The Evolution of Outcome-Based Education

Outcome-Based Education (OBE) is an innovative model for teaching and learning that aims to guarantee learners attain concrete, quantifiable results at the culmination of their educational journeys. While education systems around the globe strive to respond to changing student needs, employer expectations, and societal demands, OBE provides a systematic pathway to acquiring both academic and occupational skills (Rao, 2020).

1.1 Defining Outcome-Based Education

OBE is an educational philosophy that focuses on the clear articulation of learning outcomes, specifically the particular skills, knowledge, or abilities students are expected to acquire upon completing a course or program. In an OBE system, all elements of the educational process, from curriculum and instruction to evaluation and assessment, revolve around these pre-specified outcomes. The strategy is centered on what the learner can accomplish, rather than on what the learner is learning, so the learner's success becomes the primary gauge of achievement. OBE invites teachers to define general and discipline-specific competencies explicitly, ensuring that evaluations align with these outcomes (Syeed et al., 2022). This not only clarifies expectations for students but also lays the foundation for ongoing improvements in instructional strategies.

The major principles of OBE are:

- **Clarity of outcomes**: Measurable and clear learning outcomes direct the entire learning process.
- **Flexibility in learning paths**: Learners can attain outcomes through diverse learning methods, acknowledging different learning styles and speeds.
- **Assessment-driven instruction**: Ongoing assessment and feedback are necessary for tracking progress toward meeting the stated outcomes.

Challenges in the implementation of traditional OBE:

- **Resistance to change**: Long-established teaching methods might be firmly entrenched, making it challenging to adopt OBE structures.
- **Misalignment of curriculum**: Mapping the curriculum to align with the outcomes can be a lengthy and challenging process.
- **Lack of resources and training:** The teachers can be lacking in the tools or training needed to apply OBE successfully.
- **Complexities in assessment:** Crafting assessments that actually capture the desired outcomes can prove challenging, particularly when trying to assess complex skills.

By emphasizing outcomes, OBE ensures that students acquire practical, functional skills and abilities that will benefit them in the next phase of their professional or academic careers.

1.2 The Shift from Traditional to Outcome-Based Education

Education has traditionally been syllabus-driven, focusing on memorization and replication in exams (D'Cruz, 2017). Such methods often failed to equip students with problem-solving and application skills required in real contexts (Thirumoorthy, 2021). Outcome-Based Education (OBE) addresses these gaps by shifting the focus to competencies and real-world application. For instance, rather than recalling a formula, students are expected to apply it to practical tasks such as budgeting or measurement. This transformation influences curriculum, pedagogy, and assessment, with OBE emphasizing measurable outcomes, learner-centered instruction, and continuous feedback. Table 1 delineates the transition from content-oriented, teacher-centered approaches to outcome-based, learner-centered models that prioritize competency mastery and the practical application of knowledge (Fig. 1).

Table 1 Comparison of traditional education and outcome-based education

Aspect	Traditional Education	Outcome-Based Education (OBE)
Focus	Content delivery, syllabus coverage	Achieving specific, measurable outcomes
Assessment	Mainly summative (exams, projects)	Continuous and formative, outcome-focused
Teacher role	Content deliverer, lecture-driven	Facilitator guiding outcome achievement
Student role	Passive recipients, memorization-based	Active learners, responsible for progress
Skills	Theoretical knowledge, limited application	Application of skills in real-world contexts

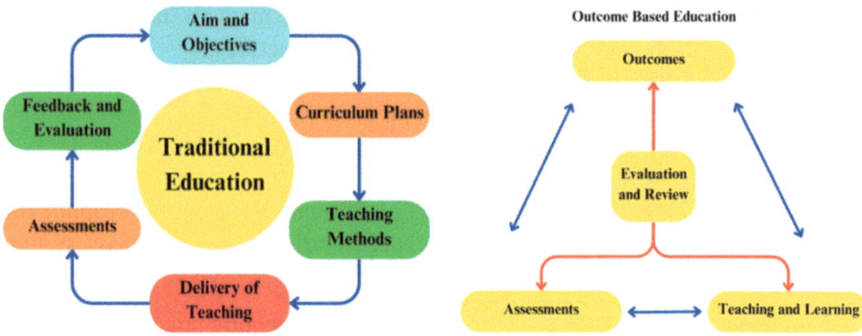

Fig. 1 Comparing traditional education with outcome-based education

1.3 *The Role of Technology in Modern Education*

The integration of technology in education is fundamentally transforming modern pedagogical practices, particularly through the widespread implementation of Outcome-Based Education (OBE). The increasing demand for performance improvements and diverse learner needs elevates expectations within educational systems, suggesting that technology may offer solutions for refining and upgrading the learning process (Deepak & Venishri, 2018).

Data-Driven Decision Making: Technology enables the ongoing collection and analysis of a significant amount of data about the students engaged in learning processes. Teachers can use this data to inform their instructional practices, track progress toward expected outcomes, and identify students who require additional support.

Adaptive Learning Systems: Adaptive learning systems are being increasingly adopted in education, driven by advancements in artificial intelligence and machine learning. Adaptive systems rely on real-time data to adjust the learning experience by tailoring the content, tests, and feedback for each learner. Adaptive systems are also ideal for the OBE model as they ensure learners achieve the expected outcomes at their own pace through personalized pathways.

Real-Time Monitoring and Feedback: New technologies enable the real-time tracking of student performance, providing both immediate feedback and opportunities for potential intervention. This is extremely valuable in an OBE environment, as it allows educators to monitor whether students are meeting specific learning outcomes and to adjust their pedagogy or provide additional materials or resources.

Collaborative and Blended Learning: Technology also supports collaborative learning contexts in the form of virtual classrooms, chat or discussion forums, and shared learning materials, making it easier for students to work collaboratively while developing essential skills such as problem-solving, teamwork, and communication. Blended learning models that combine on-campus and online learning support outcome-based education by creating additional flexibility in the manner through which students can achieve the desired predetermined outcomes.

This new technology enables the efficient distribution of content, promoting greater customization and adaptive, outcome-focused learning experiences. The latest technology enables continual assessment, quick feedback, and individualized learning, allowing both learners and educators to stay engaged in achieving competencies critical for success in today's rapidly changing world (Wang, 2024).

2 Machine Learning in Education: An Overview

Machine Learning (ML), a subset of Artificial Intelligence, is transforming the educational landscape by enabling systems to learn from data, thereby improving results without the need for explicit programming. In the backdrop of Outcome-Based Education (OBE), ML equips educators with the means to analyze student performance, customize learning experiences, and assess and confirm that learners are making acceptable progress toward achieving clearly defined competencies and program outcomes.

2.1 Machine Learning Approaches Supporting OBE

Figure 2 illustrates the primary types of machine learning approaches and their relevance to education. Each type of approach is distinctively valuable in achieving learning outcomes in Outcomes-Based Education (OBE), as they all enhance prediction or adaptive, personalized decisions.

In OBE, ML is substantial due to the availability of large datasets (e.g., student performance, test scores, and engagement) to inform better decisions and enhance learning. Machine learning can make significant contributions to ensuring that students learn in a personalized manner and achieve the stated outcomes (Panchal et al., 2025).

2.1.1 Traditional Learning in OBE

Traditional machine learning approaches, such as classification, regression, clustering, and dimensionality reduction, are essential methods for analyzing educational data within the context of OBE. Classification and regression methods can predict student success and identify students who may not successfully meet the intended outcomes. Clustering methods and pattern search can help identify students with similar learning challenges, allowing the cottage educator to develop a course of action that addresses common learning needs. Dimensionality reduction is useful for simplifying large, complex datasets, allowing educators to focus on more critical aspects of student achievement. The method, combined with traditional analysis of

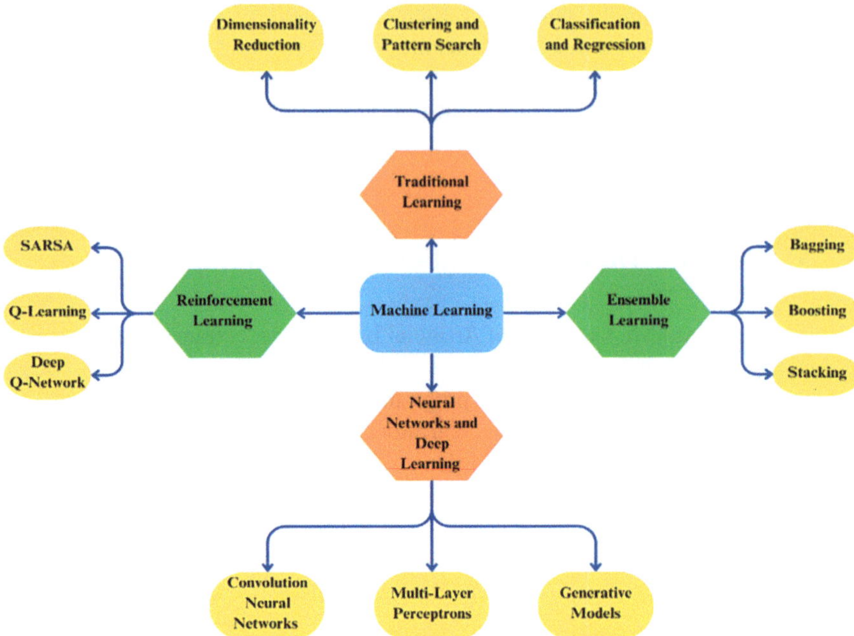

Fig. 2 The impact of machine learning on modern education

educational data aligned with data-driven teaching paradigms, supports learning activities that are linked to student achievement.

2.1.2 Reinforcement Learning in OBE

Reinforcement learning approaches, such as SARSA, Q-learning, and Deep Q-Networks, are essential to adaptive learning environments that facilitate OBE. These techniques enable systems to adaptively modify content based on the student's actions adaptively, providing adaptive feedback that increases student motivation to progress. For example, supporting students in step-wise progress toward achieving mastery of a competency will ensure ongoing development progression in coordination with achieving outcome requirements. Learning becomes adaptive, interactive, personalized, and responsive to individual needs while maintaining a focus on achieving the intended program outcomes.

2.1.3 Ensemble Learning in OBE

Ensemble learning models—such as bagging, boosting, and stacking—are a useful approach for increasing the accuracy and reliability of predictions. When it comes to OBE, these approaches enhance predictions regarding performance and assist in identifying students who are likely to need additional support earlier in the term by combining the results from multiple models. Ensemble learning reduces prediction errors in student assessments, providing a richer and more reliable source of information for educators and administrators. Ensemble learning offers several benefits, allowing educators and administrators to intervene earlier and adjust the curriculum to better align with student responses, all based on trustworthy and plausible data that meets student learning objectives.

2.1.4 Neural Networks and Deep Learning in OBE

Emerging methods for analyzing complex, high-dimensional educational data generated online include deep learning models, such as CNNs, MLPs, and GANs, as well as neural networks. Through advanced data-driven education methods, these new approaches will utilize an outcome-based education (OBE) framework to design highly personalized learning pathways tailored to individual learners. They will also support adaptive competency assessments that track the learner's ongoing real-time learning and predict future performance based on past data. This level of personalization and adaptability enables individualized support for each learner to meet program outcomes. The personalized learning will culminate in increased learner engagement and success.

Educational machine learning models can analyze large, complex datasets to produce findings that are unlikely to be uncovered by educators. This makes the educational context more streamlined, effective, and productive. Table 2 gives an overview of how various ML approaches are implemented in OBE and how integration yields benefits such as increased engagement among students, the ability to develop personalized pathways for learning, and the potential to make actionable, informed decisions based on analyzed data for improved outcomes for the students.

2.2 *Applications of Machine Learning in Education*

There is greater variability in the learning and teaching process due to the application of machine learning in education. One of the most significant applications of machine learning in education is the creation of individualized learning pathways. Learning systems that gather and analyze data on student performance histories, engagement, and learning preferences can personalize the educational experience, providing a more effective and tailored approach to learning. This depends on using learning content that aligns with students' abilities and learning styles, potentially resulting in increased engagement in learning and improved student learning outcomes. In addition to personalized pathways, intelligent tutoring systems (ITS) that utilize machine learning (ML) are enhancing the student support experience. This includes the system's ability to analyze students' knowledge and skills, provide insights into areas of struggle, and respond to students with actionable feedback and

Table 2 Applications of machine learning in education

Aspect	Relevance to OBE	Educational Application	Real-Time Example	Key Benefit
ML overview	Automates data-driven decisions	Optimizes teaching strategies	Knewton adapts content to student needs	Personalized and engaging learning
Supervised learning	Predicts performance from past data	Identifies at-risk students early	Gradescope predicts exam performance	Timely interventions, improved success rates
Unsupervised learning	Discovers hidden learning patterns	Group learners by behavior or gaps	Edmentum's exact path clustering	Differentiated instruction for varied learners
Reinforcement learning	Feedback-driven adaptive learning	Adjusts difficulty and content dynamically	Duolingo adapts lesson progression	Stepwise mastery of competencies
Data-driven insights	Identifies curriculum gaps	Analyzes trends in large datasets	Tableau dashboards for performance	Evidence-based teaching adjustments
Personalized learning	Creates tailored pathways	Customizes lessons/resources	Smart sparrow adaptive lessons	Higher engagement and satisfaction
Prediction & Decision-Making	Supports proactive academic planning	Forecasts outcomes, flags risk	Civitas learning predictive analytics	Reduced dropouts, stronger retention

individualized responses tailored to the context at the time. A notable difference for students is that this interaction is typically conducted in a one-to-one format between the student and tutor. ITS provides the ability to interact and support students in a remote format and on demand, encouraging students to be more independent and to increase their self-directed learning (Stamper et al., 2024).

Another forward-thinking implementation of predictive analytics involves using ML to predict the future success or failure of students based on data such as attendance, engagement, or study habits. These analyses may help educators identify which students are likely to struggle or fail sooner, and, using that data, can even help them initiate timely and targeted actions to promote student success. Additionally, ML is utilized to automate and enhance efficiency in tasks that consume valuable educator time, such as grading and providing feedback. For example, a type of ML, called natural language processing (NLP), assists educators in scoring free-response answers or essays, which lessens the grading workload placed on educators (Messer et al., 2023).

In an instructional setting, ML analytics can monitor student engagement in real-time and also help inform teacher adaptations in practice to maintain student engagement and identify students who may require additional support. ML can support curriculum design and development. Specifically, ML can assist teachers in making changes and improving their curriculum to better respond to students' needs by analyzing student results and comparing those outcomes with different types of pedagogical practices and materials (Alruwais & Zakariah, 2024).

2.3 Benefits and Limitations of ML for Educational Purposes

Machine learning enhances teaching in numerous ways, but it also has some issues that need to be addressed. One of the most potent aspects of machine learning in education is its ability to personalize instruction on a large scale. Unlike traditional approaches, which tend to treat all students as if they learn in the same way, machine learning can tailor the learning experience to each individual in real time. This draws students into learning and allows for alignment with an individual learner's pace of progress and preferred learning mode, thereby improving academic success. Another significant advantage of machine learning is extracting patterns from data about learners. Through the machine learning process, educators can observe how learners demonstrate trends and patterns, which informs decisions regarding curriculum development, instructional strategies, and student attention, all of which can enhance the learning experience. Another time-saving benefit of machine learning for teachers is that it can automate many of the time-consuming administrative tasks, such as grading and providing feedback on assessments, as well as scheduling. By decreasing the time spent on these regular tasks, teachers can increase instructional time and ultimately lead to better learning outcomes for students.

In addition, predictive analytics with machine learning enable the early detection of at-risk students and provide targeted interventions, differentiated support, ultimately helping to prevent academic failure. Nevertheless, some limitations should be kept in mind. Securing privacy and confidentiality is a primary concern, as the collection and use of student data must comply with both legal obligations and relevant regulations, including the General Data Protection Regulation (GDPR) and the Family Educational Rights and Privacy Act (FERPA). Securing sensitive student data correctly will help avoid significant ethical and legal challenges. Second, ML models are primarily dependent on the quality of training data. If the training data is biased against, e.g., gender, race, or socio-economic status, then the system can inadvertently reproduce the bias and generate disenfranchised data. Thus, the challenge is to ensure ML implementation is equitable and fair.

The second concern is the "black box" of specific systems, for instance, deep learning systems, that is, specific systems can prove challenging to explain or interpret (such as "why" certain decisions were taken), which can lead to transparency and accountability issues, particularly where the output of the system's recommendations has an impact on student outcomes.

ML has enormous potential to revolutionize education by enabling customization of learning, enhancing the effectiveness of teaching, and maximizing student outcomes. For it to be successful, however, attention must be given to ethical issues, data integrity, and technological limitations (Abuhassna, 2024). As we delve further into understanding how ML can be utilized for OBE, it is essential to be aware of its possibilities and pitfalls, so that it can be used responsibly and effectively in schools. The table (Table 3) discusses the most significant advantages and disadvantages of applying machine learning in education. The benefits include personalization at scale, data-driven insights, reduced administrative burdens, early detection of at-risk students, and improved learning outcomes.

3 Leveraging Machine Learning for Adaptive Curriculum Design

ML has enormous potential to revolutionize curriculum design through greater personalization, adaptability, and sensitivity to the individual needs of learners. Through the use of data-based analytics, Machine Learning has the potential to provide adaptive curricula that adapt and shift in real-time to a student's performance and engagement. The adaptive learning experience that Machine Learning provides enables learning materials to be continually adapted to meet the demands of a student's actions, ensuring that all learning materials are tailored to the individual's learning progression within the learning context. This provides educators with opportunities to modify instructional materials in relation to learners' progress rates, comprehension, and understanding of the content. For example, if a student is struggling to understand and learn a concept, the learning pathway could be amended to integrate additional practice or modify explanations to fill in any missing knowledge before the task becomes too complicated. Conversely, if a learner demonstrated mastery of a task, the system could allow the learner to continue advancing to try increasingly complex material. The ability to adapt to the individual learner results in an increasing number of learning efficiencies for the learner, as well as continues to positively engage the student's motivation to learn, ensuring that the learning materials assigned are at the appropriate level of challenge and ease.

Similarly, machine learning (ML) has an exceptional capability to review vast amounts of student data, providing ongoing feedback and helping educators identify patterns and trends that may not be readily apparent. Specifically, while examining individual students' learning behaviors, engagement, and assessments, ML can also provide feedback and actual outcomes that can be applied to update curricula, as well as improve practice and instruction. In essence, ML empowers those designing the curriculum to create an adaptive and engaging learning context that is responsive to the diverse needs of learners. This approach not only improves the effectiveness and efficiency of learning but also contributes to the advancement of a more personalized and student-centered education (Halkiopoulos & Gkintoni, 2024).

Table 3 An overview of the benefits and limitations of ML in education

Category	Benefits	Limitations
Personalization at scale	ML adjusts to the individual needs of every student in real-time, providing personalized content and learning experiences tailored to each student's unique requirements. Boosts student motivation, engagement, and academic achievement by responding to individual learning styles.	Customizing learning at scale can result in excessive dependence on technology, potentially overlooking the value of human interaction and the relationships between teachers and students.
Data-driven insights	Enables the analysis of vast amounts of student data to uncover patterns and trends.	Data may be incomplete, inaccurate, or fragmented across different institutions, limiting the effectiveness of insights derived.
	Guides decisions regarding curriculum design, teaching strategies, and student support, making education more efficient and effective.	Insights derived may sometimes be misinterpreted or lead to incorrect conclusions if not adequately contextualized within the educational environment.
Efficiency in administrative tasks	Streamlines administrative functions, such as grading, feedback, and scheduling, allowing for more time to be spent on teaching.	Overdependence on automation may lead to the dehumanization of the learning process and miscalculations in grading if systems are not constantly monitored.
	Reduces the workload of educators, allowing them to focus on more meaningful teaching interactions.	Implementing these systems may require a significant initial setup time and training for educators to fully benefit from automation.
Early identification of struggling students	Predictive models can identify students who may be at risk of falling behind, enabling early interventions.	The predictive models may not always capture the full complexity of a student's needs, leading to missed signals or unnecessary interventions.
	Helps educators provide timely, targeted support based on data such as class participation, assignment submission times, or social interactions.	Reliance on predictive data can sometimes lead to misdiagnoses of student struggles if the model is not comprehensive or is skewed by data quality issues.
Improved learning outcomes	Personalized learning with immediate feedback improves student learning outcomes.	Over-dependence on technology may result in students missing out on essential soft skills such as teamwork, communication, and critical thinking.
	Adaptive learning platforms provide alternative explanations or corrections, effectively reinforcing concepts.	The effectiveness of adaptive learning depends on the accuracy of the model's content and feedback mechanism, which may not be perfect.

(continued)

Table 3 (continued)

Category	Benefits	Limitations
Data privacy and security concerns	Collects vast amounts of data for improving personalized learning experiences.	Storing and handling personal data raises significant concerns about privacy, security, and compliance with regulations such as GDPR or FERPA.
	May lead to more effective targeting of educational interventions.	Unauthorized access or misuse of student data could lead to ethical and legal violations, damaging trust in the system.
Bias and fairness	Can tailor educational experiences to meet the needs of diverse learners, improving inclusivity.	ML models may reinforce existing biases if the training data is skewed or unrepresentative, leading to inequitable outcomes.
	Identifying and addressing gaps in student performance can help close achievement gaps.	Biases in algorithms can perpetuate gender, racial, or socioeconomic disparities, resulting in unfair treatment of certain groups of students.
Lack of transparency (black box problem)	ML systems can help optimize education processes by identifying patterns and making decisions based on data.	Many ML models, intensive learning models, are complex and lack transparency, making it difficult for educators to understand how decisions are made.
	AI can offer insights that human educators may miss, increasing overall efficiency.	A lack of explainability can lead to a lack of trust in ML systems, particularly when their decisions impact student outcomes.
Resource intensive	Provides automated solutions that can significantly increase efficiency in educational institutions.	Implementation of ML systems requires considerable computational resources, infrastructure, and ongoing maintenance, which may not be feasible for all educational institutions.
	Saves educators' time by automating time-consuming administrative tasks.	Smaller or underfunded institutions may struggle with the financial and technical demands of adopting and maintaining ML systems.
Dependence on quality data	Data analysis enables more informed decision-making and personalized learning experiences.	ML models rely heavily on high-quality, consistent data. Inaccurate, fragmented, or incomplete data can lead to flawed models and unreliable predictions.
	Quality data enable timely interventions, thereby increasing the likelihood of improved outcomes.	In many educational contexts, data may be incomplete, inconsistent, or unstructured, limiting the effectiveness of ML algorithms and predictions.

3.1 Personalization of Learning Pathways

Additionally, machine learning is heavily utilized in education by personalizing students' learning paths. Traditional curricula, more often than not, employ a linear sequence of subjects, which does not support the diverse learning needs of different students in any meaningful sense. ML-powered systems can adjust learning processes and content in line with specific learners by varying speed and material adaptively based on their learning stages, ability levels, and preferred learning styles (Widono et al., 2024).

Adaptive Learning Systems: Adaptive learning systems utilize algorithms to track a student's interactions, including quiz scores, task completion times, and engagement levels, to tailor the presentation of content. For instance, if a student is having trouble with a particular concept, the system may offer extra resources, alternative explanations, or practice exercises, and the student is not allowed to proceed until they have grasped the content.

Personalized Learning Pathways: By continuously assessing students' strengths and areas for improvement, ML models can design individualized learning pathways that enable learners to progress at their own pace. This differs from the rigid, boilerplate schedules, in which all students are assigned the same lessons to complete simultaneously. Personalizing a student's learning path enables timely intervention for at-risk students while providing opportunities for advanced students to continue making progress.

Real-Time Adjustments: Adaptive systems that utilize machine learning can make these adjustments in real-time based on a student's performance, supporting their learning within the effective zone of proximal development. An advanced learner may be assigned more complex content in a particular subject area. In contrast, a struggling learner may be given additional content or materials to support their learning. All learners are developmentally and academically engaged in the curriculum.

3.2 Predictive Analytics for Curriculum Customization

Machine learning is highly effective at absorbing information and identifying patterns within that information, facilitating an evidence-based decision-making process. For instance, in course design, ML predictive analytics can help shape course content by predicting the success of an approach to engaging different course groups (Abrorov et al., 2025).

Leveraging ML for course design enhances responsiveness, facilitates student-centered approaches, and yields meaningful outcomes for student learning. Machine learning can analyze a wide range of data, including past student performance (test data), the types of resources used, and even student demographic data. You can analyze sufficient past test performance data, combined with intervention and math reasoning, to identify students who require additional support and/or interventions. For instance, if a student is struggling with math but has average or higher performance ratings in language arts, and all other specified assessments are met, it would be reasonable to provide this student with additional support.

Curriculum designers consider these educational suggestions to spur innovation and reform; thus, the substantive reality of education with machine learning is that recognition of learning problems may be avoided, and therefore, educational policy (at all levels) can avoid a discontinuity in the students' learning process.

A well-known contribution of ML to curriculum development is in tactics, approaches, and strategies to define the effectiveness of instructional delivery. By observing student behaviors with the learning material, including in-class activities, examinations, and indicators of reading engagement, ML can assess the utility of teaching strategies for specific students. The process may reveal that one group of students, such as visual learners, is best served by instructional approaches like interactive simulations. In contrast, another group, such as auditory learners, might benefit more from podcasts or lectures. This level of understanding would enable teachers to tailor their pedagogy to meet the diverse instructional needs of students with different learning preferences, ultimately leading to improved student achievement.

With this information, curriculum planners can tailor the course structure so that instruction methods best suited to maximizing the potential of individual students are the priorities. This enables the development of an optimized and focused learning process that continually evolves based on real-time feedback.

Furthermore, ML facilitates dynamic curriculum optimization. Predictive models can constantly track and assess student performance and real-time optimize the curriculum based on emerging needs. For example, if it is statistically shown that a significant majority of students are struggling with a given subject, such as algebraic equations, then the system would suggest adding more practice exercises, breaking the content into smaller pieces, or adjusting the teaching methodologies. This adaptive curriculum type ensures that students are always challenged at the right level, avoiding the pitfalls of inflexible and uniform teaching methods. The curriculum evolves according to the students' learning paths to remain current and relevant throughout the course.

Additionally, ML will help educators identify students who are most likely to underperform or drop out of school. Through the examination of student behaviors, such as attendance inconsistencies, delayed submissions, or lack of participation, predictive analytics will identify these students to intervene and support them before harmful decisions are made. An educator may be given suggested next steps that link to more tutoring, more resources, or another assignment as the educator effectively helps the student back on track. For instance, when an ML model identifies that a student is at risk of falling behind in an online class due to low engagement and missed deadlines, the model could provide instructions to an educator or advisor to reach out to the student and share resources before the student becomes fully disengaged from the course. This proactive step enhances student outcomes and enables the teacher to identify and address potential issues or challenges that may develop into more significant problems, with implications for the student.

3.3 *Dynamic Adjustments to Teaching Methods and Content*

Using machine learning for course and curriculum design offers the exciting possibility of continually adapting the course and content in real-time based on what students are engaging with effectively or not engaging with effectively, as indicated by performance (engagement/assessment) feedback. Educators using ML can adapt not only the pace of learning but also the content of the course based on what the students are using (Dutta et al., 2024).

Active Feedback Loops: Machine learning algorithms enable educators to receive measurable feedback almost instantly, regarding each student's potential engagement with the content. For example, when students struggle with a topic in class, the ML-enabled model may recommend adjustments to instruction, such as incorporating visuals, providing more examples, or employing instructional and interactive strategies to enhance student engagement with the learning material. This type of feedback-based information can result in a timely adjustment of instruction based on data as opposed to an evaluation or survey.

Systems for Serving Data: In the same vein as video streaming services (e.g., Netflix) that suggest movies relative to a reader's preferences, ML can mediate recommendations of additional contextual materials to complement the curricular resources. For example, suppose a student is struggling with a specific concept being taught in the curriculum. In that case, the system can suggest videos, articles, or exercises that provide different verbal or visual explanations of the concept, thereby expanding the nature of the student's learning experience and resulting in increased satisfaction.

Personalized Assessments: AI can also tailor the assessment process to meet students' individual needs better. Rather than using a single assessment that works for all students, ML algorithms can administer assessments that support the needs of each learner in a differentiated manner. Those variations may be based on the difficulty level, the type of question presented, the nature of the task, or a combination of these elements. Quite naturally, this differentiation would be based on both the student's academic performance and the preferences identified by the AI system.

Gamified Learning and Engagement: ML can also augment learner engagement by determining what aspects of gamification or interactive material are most effective for individual learners. For example, suppose a specific form of game-based learning (e.g., quizzes, challenges, simulations) is found to be most effective in boosting engagement for a given set of learners. In that case, the system can dynamically add more of those aspects to the curriculum.

Scaffolding and Support: As learners progress through their tailored learning tracks, ML provides scaffolding at key moments throughout the curriculum. Support could encompass hints, instructional videos, or guided practice that decreases stepwise as the student becomes a master of the subject, allowing for more independent learning.

Through the use of machine learning in adaptive curriculum development, teachers can create highly individualized, dynamic, and responsive learning environments that cater to individual needs and enhance teaching effectiveness (Demartini et al., 2024). Machine learning's capacity for ongoing evaluation of student progress, forecasting future results, and proposing curriculum modifications renders it an influential tool for achieving the ultimate goal of OBE, enabling all students to attain the specified learning outcomes at their own pace and in accordance with their individual learning styles.

4 Real-Time Learning Monitoring: The Role of Data Analytics

In the contemporary era of education, data analysis plays a pivotal role in revolutionizing how teachers monitor, understand, and respond to student learning. As the world transitions to digital learning environments, real-time student progress monitoring has become increasingly feasible and crucial in developing responsive, adaptive, and personalized learning environments.

4.1 Tracking Student Progress with Learning Analytics

Learning analytics is the application of data to learn and improve the learning process. By monitoring students' interactions with digital learning resources, assignments, quizzes, and other course materials, teachers can gain valuable insights into how well students are progressing toward achieving their learning objectives (Vashishth et al., 2024).

Comprehensive Data Collection: Learning analytics involves gathering various data points, including quiz grades, time spent on task, discussion board participation, and social interactions in online learning settings. These data points are subsequently combined and analyzed to develop an overall picture of a student's academic performance and level of engagement.

Progress Monitoring: Learning analytics-fueled real-time tracking systems enable teachers to track students' progress toward achieving the learning outcomes defined earlier. For instance, a teacher can monitor if students are struggling with a certain subject and alter their instructional method accordingly. This also involves identifying trends in how long students take to complete an activity or when they interact most with the content.

Visualization and Dashboards: Data analytics software often displays real-time progress in the form of easy-to-understand dashboards, allowing students and teachers alike to view key performance indicators, such as completion rates, grades, and mastery of subject matter. Visualization enables teachers to immediately identify trends in student performance, making it easier to adjust teaching methods and provide targeted support where needed.

Optimization of Learning Pathways: Through constant monitoring of progress, learning analytics facilitates the optimization of customized learning pathways. For example, a student who consistently scores highly on tests may be offered higher-level materials. In contrast, a student who struggles with specific areas can be referred to supplementary materials or remediation.

4.2 Early Intervention and Support Mechanisms

One of the greatest benefits of real-time monitoring of learning is the potential to intervene early when students are at risk of falling behind. Early intervention is crucial in preventing academic failure and ensuring that students stay on track to achieve their learning objectives (Lama et al., 2024).

Predictive Models for Students at Risk: ML algorithms, when integrated with learning analytics, can identify which students will struggle based on past performance data. Late posting, non-participation, low quiz marks, or irregular attendance patterns might be precursors. By identifying such students early on, teachers can intervene promptly to fill the gaps in understanding, motivation, or other forms of learning deficits.

Individualized Support Plans: After identifying students who may be at risk, data analytics can enable teachers to develop individualized support plans tailored to each student's unique needs. Such arrangements may include additional tutoring, peer mentoring, or specially adapted learning materials designed to address one-to-one difficulties that students are experiencing. Identifying problems early on also allows teachers to provide emotional or motivational support, which, in most instances, is as essential as academic assistance.

Automated Alerts and Notifications: The majority of learning management systems (LMS) and data analysis software have automatic notification procedures that alert teachers if a student's performance or conduct is significantly below or above the expected standard. Data alerts could also prompt teachers to reach out to and connect with students, schedule functional support sessions, and/or offer encouragement and reassurance.

Proactive Curriculum Changes: Real-time tracking also enables teachers to make adjustments to instruction and the pace at which it is delivered. If a significant number of students are struggling with a specific concept, it alerts them to the fact that extra time or alternative teaching strategies are needed, so that students are not left behind in the curriculum.

4.3 *Real-Time Feedback Loops and their Impact on Student Engagement*

Real-time feedback is a crucial aspect of the learning process, as it provides students with timely insights into their performance and areas that require improvement. Learning analytics and ML facilitate the ongoing feedback cycle that has significant effects on students' motivation, engagement, and overall learning experience (Sharif & Atif, 2024).

Instant Feedback on Assessments: By incorporating immediate feedback into tests, students can receive instant assessments of their work. For instance, in an online test or quiz, immediate feedback can point out incorrect answers and provide explanations or suggestions for improvement. Not only does it help the students correct their mistakes, but it also strengthens the learning process by providing immediate reinforcement.

Gamification and Progress Indicators: Most educational platforms employ gamification techniques, such as badges, leaderboards, or milestone accomplishments, which are associated with existing performance. These systems provide students with constant feedback on their progress, as this keeps them engaged with the material. Progress indicators are used to motivate students to achieve objectives, monitor their performance, and take responsibility for the learning experience (Suartama et al., 2024).

Ongoing Engagement via Feedback: Real-time feedback ensures ongoing engagement by interfacing with the learning process and making it more interactive and responsive. When students can see their progress in real-time and receive instant recommendations for growth, they will remain engaged and motivated. On-time and pertinent feedback creates a sense of accomplishment and encourages students to keep progressing in the course.

Adaptive Feedback Mechanisms: With the support of ML, feedback is not only adaptable to the performance of the individual student but also to their preferences and learning style. For example, an average-performing student on a concept can be given additional practice questions or graphical explanations, while a higher-performing student can be given additional challenges (Mejeh & Rehm, 2024). Such a level of personalization keeps feedback useful and motivating for all learners.

Creating a Growth Mindset: Real-time feedback, particularly when it is framed in positive and supportive terms, has the potential to encourage a growth mindset among students. To continuously make students aware of their progress, give them tangible strategies, and recognize the small successes, a teacher can help students view errors as opportunities to learn information, not as failures, and this is essential to academic success over time.

Greater Accountability and Self-Regulation: Automatic feedback on student work engenders higher levels of accountability. If students know they will receive timely feedback, they will typically try harder with the course material and take responsibility for their own learning. This causes improved study behaviors, self-regulation, and ultimately improved levels of academic performance, over time.

Table 4 The role of real-time feedback loops in enhancing student engagement

Aspect	Description	Impact on Student Engagement
Real-time feedback on evaluations	Provides immediate feedback on quizzes, assignments, and tests, and alerts students to their mistakes.	Foster greater comprehension of content and increase motivation through the ability to learn from errors in real time.
Gamification and Progress indicators	Contains aspects of game-like quality, such as points, levels, or badges, and also visual cues of progress.	Boosts motivation by adding an element of success and enjoyment, motivating students to continue improving and staying focused.
Continuous engagement by feedback	Provides ongoing, regular feedback that maintains students' interest in the learning process.	Encourages learners to be actively involved in the learning process, reduces boredom and disinterest through continuous motivation.
Adaptive feedback mechanisms	Feedback adapts based on the performance of the student, providing customized tips and materials.	Strengthens personalization so that feedback is precise and applicable to specific individual student needs, leading to increased engagement.
Building a growth mindset	Stresses effort and progress over time, leading students to see problems as chances to learn.	Develops resilience, urging students to welcome learning obstacles, resulting in higher involvement and determination.
Increased accountability and self-regulation	Real-time feedback prompts students to track and evaluate their own progress, becoming responsible for their learning.	Encourages responsibility, develops self-regulation, and motivates students to take an active role in their education.

With real-time learning, data monitoring, and data analysis taken advantage of, instructors are able to create a more responsive and individualized learning environment. Keeping tabs on the progress of the students, identifying struggling learners, and providing prompt feedback not only maintains students on pace but also boosts engagement, motivation, and overall achievement. The ability for timely intervention and ongoing, actionable feedback is an empowering vehicle for the strengthening of learning performance and achieving each student's planned learning outcome (Aly, 2024). This table (Table 4) summarizes the manner in which real-time feedback loops can significantly impact student engagement by offering instant, individualized, and actionable feedback that promotes active participation and a growth mindset.

5 Research Gaps in AI-Driven Outcome-Based Education

Notwithstanding the increasing potential of ML in informing OBE, some key research gaps must be filled to ensure the full exploitation of the potential of AI in educational environments. These discrepancies range across different dimensions, from scalability and interoperability with current education systems to ethical issues and long-term impacts on teaching and learning. Such gaps must be filled to ensure the successful, fair, and sustainable deployment of AI-based OBE systems (Gnana Rajesh et al., 2024).

5.1 Scalability and Generalization of ML Models

One of the biggest challenges to applying ML to outcome-based education is the limitations on the scalability and generalizability of ML models. While ML algorithms have been shown to work well in small or controlled learning environments, the extent to which ML works well in larger heterogeneous learning environments is unknown.

For ML models to be favorably embraced by school systems, they should be generalizable across a range of types of institutions, classes, and student populations. Future studies on ML in education must aim at creating a concept of the model's flexibility across learning settings and in a way that does not compromise their sophisticated individualized personalization abilities. Future studies should also investigate how various populations of students defined by diversity in preexisting knowledge, motivation, and learning requirements can interact with adaptive learning environments. Scalability and sustainability of ML systems without compromising the capacity to deliver individualized learning experiences is the premise for competing and maintaining any long-term sustainability of ML in general in education.

5.2 Integration with Current Educational Paradigms

The integration of AI-powered tools within dated educational frameworks is yet another central challenge. Schools function mostly through tried-and-proven models of instruction, which might not have the capacity to fully support the adaptive, data-driven characteristics of ML. Welcoming ML into these frameworks very easily destabilizes venerable approaches to instruction and curricula.

There must be research to assist us in gaining a useful understanding of how best to operationalize ML on these legacy systems. This can involve looking into hybrid models that move between legacy instructional and ML-based adaptive learning trajectories. There also must be research to determine what curricular and pedagogical changes would be required in order to incorporate ML as smoothly and effectively as possible for students and teachers. The incorporation of ML into existing systems to peer into the future will also consider the technology implications on pedagogy and effective teaching practice.

5.3 *Long-Term Impact and Adaptation of AI-Driven OBE*

Although the short-term benefits of ML in OBE are evident—improved student engagement and performance—we know much less about the long-term effects of the technology-enabled OBE systems using AI.

To comprehensively explore the role of artificial intelligence in education, longitudinal studies that study the effect of machine learning on learning sequences, retention, and growth in education over time across several years are needed. The study would evaluate the immediate effects of AI on educational outcomes, for example, learning with AI, but would also evaluate how student learners sustained the development of skills and outcomes in cognition over time, i.e., cognitive outcomes over time. It would also explore what instructional practices of teachers are in relation to AI, and how that affects teacher style and approach to learning, and how they respond to AI-enabled tools in their instruction. The proposal represents a significant avenue for explicating how teaching practices evolve with AI-influenced teaching and learning practices, and if they have an impact or produce an impactful, transformative experience for students capable of long-term retention.

5.4 *Ethical Implications and Bias in Data*

A significant concern in the application of machine learning (ML) in education is the possibility of data bias being used to train an AI system. However, real-world machine learning algorithms have the potential for bias based on the data, which can subsequently bias the outcomes that affect students. Bias, in this context, refers to inequities associated with characteristics such as race, gender, and socio-economic status. If the bias is perpetuated, it can further exacerbate existing inequities in education.

Social equity is a call for research on algorithmic bias to determine how bias can be defined, identified, and lessened, and ideally eliminated, from biasing data feeding AI systems. This involves developing procedures to ensure the data used to train ML models is representative and does not contain bias. Ethical considerations must also be explicitly considered when designing AIBOBE learning systems. Research must explore the potential outcomes of AI decision-making in education, particularly with regard to issues such as grading and personalized learning streams. To maintain trust and acceptability of AI technology, ways need to be found to demonstrate fairness and justice in AI usage.

5.5 Teacher Training and Professional Development

The proper knowledge and skills to implement these technologies in a classroom must be held by teachers. In general, most teachers do not have significant professional development experience to know and use machine learning tools in their own teaching context. Research is needed to design and deliver professional learning experiences that go beyond simply showing teachers how to interface with AI tools, and allow for the intended practice to imagine their pedagogical practice using the AI potential and learning tool. While professional learning is important to be technical, educators need to see AI used to enhance and deepen pedagogy and ultimately, student outcomes. Professional learning experiences will position educators to strive to use machine learning as a legitimate pedagogy of learning and to use student data to create meaningful learning experiences.

5.6 Student Privacy and Data Security

The implementation of machine learning (ML) technologies in educational contexts almost always requires significant amounts of student data to be collected and curated, but these steps raise serious concerns about student privacy and the safeguarding of, or a rush to safeguard, educationally sensitive information. In addition, privacy and misuse are of increasing public awareness and concern. As a result, the magnitude of the issue is growing.

Nevertheless, in creating AI-based outcome-based education (OBE) platforms, further effort is required to offer future research agendas for designing privacy-guaranteeing frameworks for the utilization of student data. The research agenda should encompass examining the anonymization methods for student data, privacy policies, and compliance challenges, and encryption methods for sensitive student data protection. The discussion/research that provides solutions to overcoming the difficulty in balancing between the extensive data needs for efficient AI learning uses and the privacy of students should be ongoing. Without data protection/security and privacy in data, the application of AI in learning ethically and responsibly is excluded.

It is necessary to tackle research opportunities to fulfill the maximum potential of AI in OBE. Along with the growth of ed-tech eco-systems, it is essential to address the issues of scalability, integration, long-term impacts, bias, teacher training, and data privacy and security in order to be able to create effective, accessible, and sustainable AI-assisted education systems. With emphasis on these aspects, AI-enabled OBE systems will not only improve the learning experience but also facilitate inclusive, accessible, and equitable education (Saxena et al., 2024).

6 Future Directions: The Impact of Machine Learning on the Future of OBE

With the ongoing development of ML technologies, they will redefine the future of OBE by stimulating innovations in curriculum planning, evaluation, and pedagogical practice (Prihantoro, 2023).

6.1 The Evolution of Curriculum Design

The planning of curriculum is on the cusp of a transformative revolution driven by the expanding utilization of machine learning. Curricula in the past have been created using a one-size-fits-all approach, with little scope for differentiation to accommodate diverse learning requirements, interests, and abilities of students. Machine learning has the ability to develop knowledgeable and student-centered curricula that acknowledge individual learning paths and styles while allowing for a more flexible approach than static and universal approaches.

Dynamic Curriculum Development: Future curricula will not be a static document. Instead, a dynamic curriculum framework based on data will be responsive and adapt to the classroom environment in real-time based on student progress. Machine learning models will continuously evaluate data such as student participation, assessment scores, and learning behaviors to optimize how the material is presented to students. For example, if the majority of students are struggling with a specific topic or subject, the curriculum will update automatically and allocate additional time to that content or provide more varied learning resources to help students better understand. Resources might include videos, articles, or interactive movement and exercise when appropriate.

Personalized Learning Progressions: Machine learning will generate the next-level curriculum personalization and essentially create individualized learning pathways for students based on their strengths, weaknesses, and interests. Recommendations will be based on analyzing patterns from students' performance data. Machine learning will even determine and generate customized content, suggest the optimal order to learn from the learning materials, or recommend extra activities to the students. A profound level of personalization has the potential to render the educational experience more effective and more interesting, particularly for students who excel in individualized experiences as compared to a conventional or standard one-size-fits-all mode.

Dynamic Content Adaptation: In addition to machine learning algorithms and models that monitor students' progress, the algorithms will also possess adaptive features to suggest instant or real-time revision of the content, ensuring that learning paths are continuously adapted, as well as in real-time, to individual students' needs as they change. For example, if a student demonstrates mastery of a specific skill, the curriculum may transition them to advanced-level content. Alternatively, if a learner is struggling, the system can adjust in real-time to deliver foundational content or additional practice problems (Arun Kumar, 2023). Not only does this responsiveness ensure that all learners progress at an appropriate pace, but it also reduces the risk of learners becoming disengaged or overwhelmed by material that is either too challenging or too easy.

Collaborative and Interdisciplinary Learning: Machine learning can also enable the development of curricula that promote interdisciplinary learning and collaboration. Through monitoring a student's performance in several subjects, ML algorithms can detect potential areas of overlap between subjects and recommend collaborative assignments or projects that cut across disciplines. This approach fosters critical thinking, creativity, and problem-solving skills, which are essential in today's world.

6.2 AI-Powered Assessment Models

OBE relies on assessments as its core component because assessments can measure the extent to which students have achieved the learning outcomes. Conventional models of assessment often overlook a significant portion of a student's comprehensive learning experience and can be biased, rigid, or overly rote-oriented (Rani et al., 2024). ML can revolutionize assessment through more adaptive, thorough, and personalized evaluation systems beyond traditional test-taking.

Continuous Formative Assessment: Perhaps the most exciting use of ML in assessments is the move away from summative, end-of-term tests to continuous formative assessments. ML-driven systems can monitor students' progress in real-time, monitoring their participation, engagement, and performance across different tasks. Rather than relying on a single end-of-course test, students will be assessed continuously through small, frequent activities that provide ongoing feedback on their learning journey. This enables teachers to identify and address issues related to decisions about student learning. This is an important skill to ensure that educators can anchor their judgment and actions. AI-based evaluation models will not only provide personalized assessments but will also present real-time, customized feedback. For example, an AI system could analyze a student's response to an essay or project and provide instant feedback on issues such as clarity, rationale, or evidence. That feedback may be paired with suggestions for how to do better and citations of resources that provide support, so students can correct their work step by step. This personalized, immediate feedback makes learning more intense and encourages students to work more intensively on the material.

Adaptive Testing: In the classroom of the future, AI-driven tests can incorporate adaptive testing strategies, where the level of difficulty in questions dynamically adjusts based on a student's performance. For instance, if a student answers a set of questions correctly, the system can increase the level of difficulty; however, if a student is found to be struggling, the system can decrease the number of questions.

Skills-Based and Competency-Based Tests: Machine learning will also facilitate the transition to competency-based and skills-based testing, where students are assessed not only on their knowledge of facts but also on their ability to apply that knowledge to accomplish tasks in real-world settings. ML models will be trained on a variety of student data, including projects, simulations, and peer reviews, to assess their proficiency level in numerous skills (Minalla, 2024). The method ensures that the assessments measure deeper learning and better evaluate a student's capability to achieve the desired learning outcomes.

AI in Peer and Self-Assessment: Next-generation AI-based systems can also facilitate peer and self-assessment, where students evaluate their own work and that of their peers. Such methods can be supplemented by machine learning algorithms, which can provide guidance and templates for assessing quality, offering suggestions for improvements, and even detecting patterns of bias or discriminatory grading in peer review. This can facilitate critical thinking and reflection, which are often neglected in traditional assessment.

This table (Table 5) highlights how AI-based assessment models can transform traditional assessment procedures by making them continuous, personalized, and responsive, while enhancing the accuracy and efficiency of measuring students' competencies.

Table 5 Benefits of AI-powered assessment models in education

Aspect	Description	Impact on Student Learning
Continuous formative assessment	AI continuously tracks student performance, providing frequent checkpoints rather than relying solely on summative tests.	Facilitates continuous learning by providing students with timely feedback on their performance, enabling them to make more effective learning adjustments.
Automated and personalized feedback	Artificial intelligence programs evaluate student assignments and provide customized feedback, highlighting areas of strength.	Saves teachers' time and provides students with immediate, specific feedback that promotes personalized learning and understanding.
Adaptive testing	AI dynamically adjusts the difficulty level of tests based on student performance, ensuring the test is challenging enough to meet individual needs.	Guarantees that students are constantly being challenged at the right level, thereby enhancing engagement and facilitating reflection on progress.
Skills-based and competency-based assessments	AI assesses whether students have acquired specific skills or competencies, rather than relying solely on grades.	Gives a more accurate representation of student competency, enabling personalized learning streams and greater alignment with skills in the real world.
AI in peer and self-assessment	Artificial intelligence tools help students review their own work and that of others, providing formal frameworks for feedback.	Promotes peer learning and self-reflection, enhancing critical thinking and collaborative abilities.

6.3 Preparing Educators for the Future Classroom

For machine learning and cutting-edge technologies to merge into OBE, educators must be adequately trained to apply these technologies in the classroom. Teachers need to prepare for a future where ML becomes integrated in the school through a paradigm shift in teacher training, professional development, and support systems (Kirana et al., 2024).

Teacher Education in Data Literacy: Teachers shall have to be trained in data literacy so that they can adequately comprehend and use the insights developed by ML technologies. This implies not only knowing how to read data but also using it to inform teaching practice. The teacher training would need to encompass the acquisition of data analysis, learning analytics applications, and how ML-facilitated systems can be used to change pedagogical methods and reconfigure student learning pathways.

Development of a Growth Mindset among Teachers: With technology integration in schools at an all-time high, teachers need to possess a growth mindset and be open to trying new tools, continue refining their practices, and learn from errors when implementing technology. Schools will need to support their teachers through this transition by allocating time for mentorship and professional development, as well as providing the necessary time and space for collaborative teaming. This will help educators gain confidence in working with ML tools and effectively employ them to improve students' performance.

Ethical Decision Making and Training: Beyond content knowledge, instructors will also require training in the moral implications of deploying machine learning in the educational sector, including considerations of privacy, security, and equity. As educators deploy algorithms into their practice to make more informed student learning decisions, this is an important skill to ensure that educators are able to anchor their judgment and action in algorithms. Teachers will need some general technical education, along with a broader understanding of social values such as equity and inclusion, to be suitable and effective in employing algorithms.

Teacher-Data Scientist Collaboration: Data scientist, technologist, instructional designer, and teacher collaboration will be required for sustainable data usage of ML in OBE. This continued collaboration will ensure that ML tools are well-integrated and utilized for teaching and learning outcomes in continuous curriculum development. Interdisciplinary teams of these experts may collaborate to enhance learning goals, provide meaningful feedback, and accommodate the unique characteristics of a particular classroom. Regular collaboration will also habituate teachers to work with these technologies, making the tools practical and applicable to their practice.

Creating a Community of Practice: Part of preparing teachers for the future classroom involves developing communities of practice where teachers can share knowledge, successes, and concerns about using machine learning in the school. Professional learning networks (PLNs) and online collaborative platforms will provide teachers with opportunities to stay current with best practices, exchange resources, and connect with educational technology experts.

The future of OBE is inextricably intertwined with the continuous advancement of Technologies. The development of curriculum design, the emergence of AI-driven assessments, and the readiness of teachers for future classrooms will determine how education systems adapt to serve the needs of a heterogeneous, rapidly evolving student body (Asbari & Novitasari, 2024).

7 Practical Roadmap for Implementing Machine Learning in OBE

This chart (Table 6) presents a step-by-step action plan for incorporating machine learning into OBE, divided into ten significant steps. It begins with identifying learning goals and mapping them against the advantages of ML, followed by evaluating the existing technological infrastructure to determine if it is ready for integration. The steps involve selecting the right AI tools, testing ML-led projects, and educating teachers to utilize the technology effectively (Aminah et al., 2025).

ML tools are then scaled throughout the curriculum, with ongoing observation of their effect on learning outcomes. Ethical and privacy issues are resolved, followed by scaling the implementation throughout departments. Lastly, the roadmap focuses on fostering ongoing innovation to maintain pace with emerging AI technologies, ensuring constant improvement and success in ML-based education (Fig. 3).

8 Conclusion: Reimagining Education Through Machine Learning

The incorporation of ML into outcome-based learning is a significant step toward innovations in education that respond to the diverse and dynamic needs of learners. With adaptive learning routes, real-time monitoring, and tailored evaluations of learning, ML can potentially support a shift to the learners' study and teachers' instruction. The current chapter outlines the primary contribution of ML to overcoming issues in conventional learning, offering personalized learning experiences, and enhancing students' performance. Nonetheless, with promise comes significant challenge, especially in the domains of ethics, privacy, and scalability for ML in OBE, as it seeks to fulfill its promises. Overcoming these challenges is necessary to ensure the ethical, equitable, and sustainable deployment of AIs in various educational environments.

Table 6 Step-by-step guide to integrating machine learning in OBE

Step	Objective	Actions	Practical Applications
Step 1	Identify key educational objectives and outcomes	Define desired learning outcomes and educational goals.	Utilize machine learning to analyze historical student performance, aligning learning outcomes with student needs and course expectations.
Step 2	Assess current technological infrastructure	Review current learning management systems (LMS), data storage solutions, and computational resources.	Use tools like Google classroom analytics or Moodle to assess current LMS capacity, bandwidth, and data processing capabilities for ML integration.
Step 3	Select appropriate AI tools and platforms	Identify ML platforms with features such as personalized learning pathways, predictive analytics, and real-time feedback.	Consider solutions like Knewton for adaptive learning or IBM Watson for AI-driven learning, ensuring they are seamlessly integrated with existing systems such as blackboard or canvas.
Step 4	Pilot ML-driven OBE initiatives	Choose a group of students or courses to pilot.	Deploy an ML-driven adaptive learning software (e.g., squirrel AI) within a particular class, and monitor its effect on participation and learning outcomes via dashboards.
Step 5	Educate educators and staff on integrating AI	Create specialized training programs for teachers.	Implement online training and certification programs, such as Coursera's AI for educators, to reskill teachers to leverage AI-driven tools and methodologies in their classrooms.
Step 6	Incorporate AI into the full curriculum	Apply pilot results feedback to improve the integration of ML tools throughout the curriculum.	Utilize platforms like Smart sparrow to tailor curricula, content, and assessments to the diverse needs of students across various disciplines.
Step 7	Monitor and evaluate impact	Set KPIs for tracking students' progress, active participation, and grades.	Implement real-time monitoring systems, such as Google analytics for education, to assess the impact of ML-based interventions on student engagement and performance.
Step 8	Address ethical and privacy concerns	Enact data privacy policies in accordance with relevant laws, such as the GDPR and FERPA.	Implement encryption software such as Bitdefender or AWS encryption services to secure student data and ensure that all data gathered is in conformity with privacy legislation.

(continued)

Table 6 (continued)

Step	Objective	Actions	Practical Applications
Step 9	Scale and expand the ML implementation	Broaden the coverage of ML tools to other courses, departments, and student populations.	Scale the ML infrastructure using cloud services like AWS or azure, enabling more students and courses to be supported without compromising performance.
Step 10	Foster continuous innovation	Promote continuous research into AI and education technology advancements.	Develop an innovation center or lab where teachers can test new technologies such as AR/VR, AI simulations, or gamification platforms for customized learning.

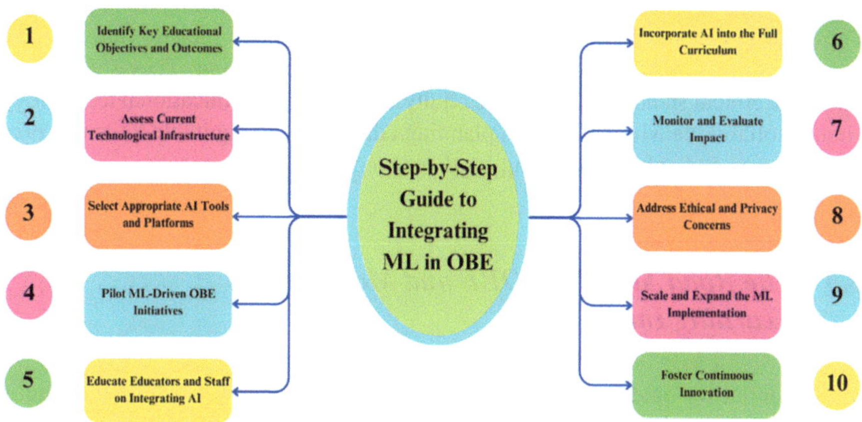

Fig. 3 Step-by-step guide to integrating machine learning in OBE

8.1 *Summarizing Key Insights*

Machine learning can enhance outcome-based education (OBE) by enhancing personalization, making more accurate predictions of student performance, and facilitating real-time adjustments to curricula, which in turn enhances involvement and improves educational outcomes, especially in multicultural and multiracial classrooms. By facilitating ongoing analysis, ML adds another level of continuous monitoring and early intervention, resulting in improved retention and student performance. While systems that integrate ML offer benefits, there are also challenges, including data privacy, algorithmic bias, and equity of access to technology. The principles of equity, transparency, and ethics should also inform implementation. Support needs to be viewed as a continuous professional learning process, not based on the utilization of tools, but on the pedagogical implications of AI and how it is applied. When an ML-driven OBE system scales, significant challenges emerge. Generic, flexible solutions are required that can be used across different institutional contexts and reconfigured to fit their specific structures. Flexibility, inclusivity, and deliberate intent are integral considerations for effective and sustainable AI OBE solutions.

8.2 Practical Recommendations for Stakeholders

For outcome-based education via machine learning to be successful, teachers need to be prepared and supported. Schools should be capable of piloting small-scale experiments to test how ML can be implemented within their classrooms, and with a reasonable investment in infrastructure expenses, while safeguarding student data. Teachers require proper time, technical support, and experience collaborating. Students should assist in planning how ML will make their learning more personal. Policymakers should strive for equitable access to ML-based tools for all students, regardless of the school they attend or their level of resourcing. They should also influence ethical standards to advance equity. Developers of educational technology should collaborate with teachers in planning adaptable and equitable tools that promote privacy and enhance learning for diverse educational approaches.

8.3 The Road Ahead for OBE and Machine Learning Integration

The potential for outcomes-based education (OBE) driven by machine learning is fascinating, and it is imperative to apply design thoughtfully and ethically.

To use ML in OBE education optimally, institutions can consider:

- **Scalability**: Ensuring that ML models can scale and be adaptable to work well in different education systems, curricula, and with diverse students.
- **Equity**: Addressing issues of access to technology and offering AI-based learning streams to every student, regardless of socio-economic background.
- **Teacher Empowerment**: Training teachers on what they should know to leverage ML to offer individualized and effective teaching.
- **Continuous Improvement:** As a process, utilizing AI in OBE is continuous and is required to be constantly reviewed, analyzed, and updated. Institutions must foster an environment of innovation and regularly monitor emerging AI opportunities to explore the possibility of adoption.
- **Ethical Implications:** Along with machine learning becoming more pervasive in the education sector, it is equally important to heed ethical concerns as well. Schools need to ensure that their ML models are safeguarding data, being transparent regarding decisions by algorithms, and ensuring fairness.

The ML methodology, in conjunction with outcome-based education, can lead to a more efficient, adaptable, and personalized learning process. If we can address the noted considerations, we can promote an environment where all concerned parties can engage towards a balanced, just, and functional future, where students of all levels will reap the rewards of OBE learning, as hypothesized based on the perspective of machine learning and AI.

References

Abrorov, A. et al. (2025). *Optimizing mechanical engineering curriculum design using AI and machine learning*. AIP Conference Proceedings, 3268, p. 070036. https://doi.org/10.1063/5.0258030.

Abuhassna, H. (2024). The information age for education via artificial intelligence and machine learning: A bibliometric and systematic literature analysis. *International Journal of Information and Education Technology, 14*(5), 700–711. https://doi.org/10.18178/ijiet.2024.14.5.2095

Alruwais, N. M., & Zakariah, M. (2024). Student recognition and activity monitoring in E-classes using deep learning in higher education. *IEEE Access, 12*, 66110–66128. https://doi.org/10.1109/access.2024.3354981

Aly, M. (2024). Revolutionizing online education: Advanced facial expression recognition for real-time student progress tracking via deep learning model. *Multimedia Tools and Applications*. https://doi.org/10.1007/s11042-024-19392-5

Aminah, S., Krisnadhi, A. A., & Hidayanto, A. N. (2025). Ontological framework for the analysis of outcome-based curriculum in higher education. *IEEE Access, 13*, 31497–31516. https://doi.org/10.1109/access.2025.3542881

Arun Kumar B. R (2023) *Outcome-based education model*. Handbook of outcome-based education model (OBE): Best Practices and Formats for OBE Implementation in Technical Higher Education Programs. 2–14. doi:https://doi.org/10.9734/bpi/mono/978-81-19315-93-2/ch1.

Asbari, M., & Novitasari, D. (2024). Outcome-based education model: Its impact and implications for lecturer creativity and innovation in higher education. *International Journal of Social and Management Studies, 5*(5), 22–31. https://doi.org/10.5555/ijosmas.v5i5.447

D'cruz, L. (2017). Transition from traditional to outcome-based education–a case study. *IOSR Journal of Research & Method in Education (IOSRJRME), 07*(04), 38–45. https://doi.org/10.9790/7388-0704013845

Deepak, T. J., & Venishri, P. (2018). Outcome based education OBE trend review in engineering education. *International Journal of Trend in Scientific Research and Development*, (Special Issue-ICAEIT2017), 41–43. https://doi.org/10.31142/ijtsrd19126

Demartini, C. G., et al. (2024). Artificial intelligence bringing improvements to adaptive learning in education: A case study. *Sustainability, 16*(3), 1347. https://doi.org/10.3390/su16031347

Dutta, S. et al. (2024) *Enhancing educational adaptability: A review and analysis of AI-Driven adaptive learning platforms*. 4th International Conference on Innovative Practices in Technology and Management (ICIPTM), 1–5. https://doi.org/10.1109/iciptm59628.2024.10563448.

Gnana Rajesh, D., Dheva Rajan, S., & Punnolli, M. (2024). *Strategic perspectives on integrating AI into activity-based learning: a pedagogical approach*. 5th International Conference on Image Processing and Capsule Networks (ICIPCN), 226–232. https://doi.org/10.1109/icipcn63822.2024.00044.

Halkiopoulos, C., & Gkintoni, E. (2024). Leveraging AI in E-learning: Personalized learning and adaptive assessment through cognitive neuropsychology—A systematic analysis. *Electronics, 13*(18), 3762. https://doi.org/10.3390/electronics13183762

Kirana, N. P., Lintangsari, A. P., & Emaliana, I. (2024). Preparing EFL pre-service teachers for inclusive teaching: The students' perception on the outcome-based evaluation of the introduction to inclusive education course. *English Language Teaching Educational Journal, 7*(1), 43–54. https://doi.org/10.12928/eltej.v7i1.10280

Lama, A. V., et al. (2024). Revolutionary teaching approaches. *Journal of Information Systems and Technology Research, 3*(2), 73–79. https://doi.org/10.55537/jistr.v3i2.829

Mejeh, M., & Rehm, M. (2024). Taking adaptive learning in educational settings to the next level: Leveraging natural language processing for improved personalization. *Educational Technology Research and Development, 72*(3), 1597–1621. https://doi.org/10.1007/s11423-024-10345-1

Messer, M., et al. (2023). Automated grading and feedback tools for programming education: A systematic review. *Acm Transactions On Computing Education, 24*(1), 1–43. https://doi.org/10.1145/3636515

Minalla, A. (2024). Competency-based learning for improved teaching and learning the thermodynamics course for chemical engineering students. *ASEAN Journal of Engineering Education, 8*(2), 127–135. https://doi.org/10.11113/ajee2024.8n2.168

Panchal, D., Champaneria, M., Agarwal, A., & Shah, M. (2025). Unlocking the power of machine learning in education: A comprehensive overview of opportunities and challenges. In *CRC Press eBooks* (pp. 348–368).

Prihantoro, C. R. (2023). Outcome Based Education (OBE) based vocational education model in the era of Artificial intelligence (AI). *Education Quarterly Reviews, 6*(1). https://doi.org/10.31014/aior.1993.06.01.701

Rani, S., Kaur, G., & Dutta, S. (2024). Educational AI tools: A new revolution in outcome-based education. In *Explainable AI for Education: Recent Trends and Challenges* (pp. 43–60). https://doi.org/10.1007/978-3-031-72410-7_3

Rao, N. J. (2020). Outcome-based education: An outline. *Higher Education for the Future, 7*(1), 5–21. https://doi.org/10.1177/2347631119886418

Saxena, N. K., Bansal, R. C., & Saxena, V. (2024). *Machine learning based strategy for students performance improvement*. 13th International Conference on System Modeling & Advancement in Research Trends (SMART), 621–626. https://doi.org/10.1109/smart63812.2024.10882538.

Sharif, H., & Atif, A. (2024). The evolving classroom: How learning analytics is shaping the future of education and feedback mechanisms. *Education Sciences, 14*(2), 176. https://doi.org/10.3390/educsci14020176

Stamper, J., Xiao, R., & Hou, X. (2024). *Enhancing LLM-based feedback: Insights from intelligent tutoring systems and the learning sciences*. arXiv (Cornell University) [Preprint]. https://doi.org/10.48550/arxiv.2405.04645.

Suartama, I. K., et al. (2024). Student engagement and academic achievement: The effect of gamification on case and project-based online learning. *Journal of Education and Learning (EduLearn), 18*(3), 968–982. https://doi.org/10.11591/edulearn.v18i3.21349

Syeed, M. M., et al. (2022). Outcome-based education (OBE): Defining the process and practice for engineering education. *IEEE Access, 10*, 119170–119192. https://doi.org/10.1109/access.2022.3219477

Thirumoorthy, G. (2021). Outcome-based education (OBE) is the need of the hour. *Educational Quest- an International Journal of Education and Applied Social Sciences, 12*(1). https://doi.org/10.30954/2230-7311.1.2021.6

Vashishth, T. K., et al. (2024). AI-Driven Learning Analytics for personalized feedback and assessment in higher education. In *Advances in media, entertainment and the arts (AMEA) book series* (pp. 206–230). https://doi.org/10.4018/979-8-3693-0639-0.ch009

Wang, L. (2024). Reform and practice of College English Reading teaching based on OBE concept. *Frontiers in Educational Research, 7*(5). https://doi.org/10.25236/fer.2024.070531

Widono, S. et al. (2024). *A strategic design of personalized based learning system for improving the experience of outcome based education*. 2024 4th International Conference on Advance Computing and Innovative Technologies in Engineering (ICACITE), 1149–1154. https://doi.org/10.1109/icacite60783.2024.10616811.

Exploring Teachers' Perceptions of Machine Learning in Outcome-Based Education

Muhammad Arif, Hira Fatima, Muhammad Aqib, and Miracle Uzochukwu Okafor

Abstract This study examines teachers' perceptions of Machine Learning (ML) tools within the context of Outcome-Based Education (OBE) to understand how these technologies are influencing instructional practices in contemporary classrooms. As educational systems evolve to meet the demands of the twenty-first century, ML offers transformative potential through personalized learning pathways, adaptive assessments, and real-time performance analytics. Using a qualitative approach, this study investigated how educators perceive the effectiveness, usability, and pedagogical alignment of ML tools. In-depth interviews, focus groups, and classroom observations were conducted to identify the perceived benefits and challenges faced by teachers in implementing the program. Key barriers include limited technical proficiency, difficulties with system integration, ethical concerns related to data privacy, and uncertainty surrounding the use of algorithmic decision-making. These findings highlight the critical role of professional development and institutional support in facilitating effective ML integration. Ultimately, this study contributes to the growing body of literature on AI in education by offering insights into how teacher readiness and perceptions influence the successful adoption of ML tools within the OBE framework.

Keywords Machine learning · Outcome-based education · Teacher perceptions · Educational technology · Qualitative research · Technology integration

---

M. Arif (✉)
School of History and Culture, Southwest University, Chongqing, China
e-mail: arifedu8@swu.edu.cn

H. Fatima
Faculty of Education, Southwest University, Chongqing, China

M. Aqib
College of Computer Science, Taiyuan University of Technology, Taiyuan, China

M. U. Okafor
School of Foreign Languages, Hanjiang Normal University, Hubei, China
e-mail: mu.okafor@unizik.edu.ng

D. Vetrithangam et al. (eds.), *Transforming Outcome-Based Education with Machine Learning*, https://doi.org/10.1007/978-981-95-4925-2_3

1 Introduction

Over the last twenty years, Machine Learning (ML) has evolved from a laboratory topic to a practical application that teachers and institutions actually utilize (Arif et al., 2025a; Anis, 2024). Initially, researchers aimed to automate paperwork and develop adaptive lessons. With faster computers, more data, and better algorithms, ML now supports many everyday tasks across education (Dahlin, 2021; Zamiri & Esmaeili, 2024). Modern educational settings use such platforms for individualized teaching methods, student forecast modeling, and test enhancement (Taylor et al., 2024). Outcome-Based Education (OBE) has established itself as an educational framework that focuses on defining clear student achievements and constant improvement in teaching. These insights share their ability to generate specific learning outcomes through data-driven education insights with OBE while achieving the goal of clear educational outcomes (Song et al., 2024). This analytical information supports academic institutions in identifying practical pedagogical methods and enables them to modify their teaching approaches while implementing systemic education reforms promptly.

Developing these technologies brings two critical benefits to educational institutions, as they can actively monitor teaching practices while delivering personalized learning experiences, according to Rashid and Kausik (2024). These tools enable teachers to receive ongoing feedback on their teaching methods by allowing them to track performance and predict analytic data (Dwivedi, 2022). The input provided enables educators to observe patterns in student response behavior, which improves educational results. The advancement of adaptive systems aligns perfectly with OBE, since these educational approaches share the key concept that continuous improvement leads to defined educational benchmarks (Sivarajah et al., 2021). The implementation of ML within OBE frameworks provides a way for teaching practices to use evidence-based methods, rather than standard exam ratings.

The growing number of publications on ML applications in education lacks sufficient analysis of teachers' perspectives on these technological developments (Ming & Li, 2025). Technology implementation success depends heavily on the acceptance of teachers who remain active participants rather than passive end-users of these systems because their readiness and perception directly influence the results. Existing studies have mainly concentrated on technological advancements and student performance outcomes without properly considering human factors. According to Davis (1989), user acceptance remains the fundamental determinant for technology adoption within schools, where teachers' perception levels directly impact how well they use and adopt ML tools.

Current studies reveal that teachers encounter three main barriers: inadequacy of technical skills, resistance to change, and worries about data privacy (Arif et al., 2025b; Pusporini & Nurdiyanto, 2024; Maghsudi et al., 2021). The successful implementation of ML in OBE requires an investigation of teachers' perspectives on the factors that determine its success. This study required both benefit analysis and barrier identification to prevent technological obstacles in teaching practice transformation (Kour et al., 2021). Educational outcomes and their relationship with pedagogical innovation and technology acceptance are the focus of this study because they approach the analysis from the perspective of teachers' perceptions.

This study asks a simple question with big implications: How do teachers actually experience ML tools in OBE settings? To answer it, we probe six areas: how effective these tools are for teaching quality and student outcomes; how well they support personalization and student engagement; what barriers teachers face and the workarounds they use; which ethical concerns matter most—privacy, bias, and transparency; how readiness factors such as perceived usefulness and perceived ease of use play out in real workplaces; and what kinds of professional development and institutional support teachers say they truly need.

The primary objective of this research was to obtain personal experiences and knowledge of the teachers about ML tools within the OBE learning systems. The results are both theoretically and practically relevant to the improvement of the use of ML in the learning environment. This study will enhance the discursive map of educational technology by implementing both theoretical frameworks, the Technology Acceptance Model (TAM) and the Constructivist Learning Theory (CLT), towards the evaluation of the perception as understood by the teachers. This study combines these theoretical frameworks in order to comprehend the relevance and convenience of the use of the TAM constructs by the teachers, along with the pedagogical aspects of scaffold and learner-centered practices in the light of CLT (Vygotsky, 1978). The proposed new research method combines various constructs to learn about the way teachers react to ML tools that enhance already existing knowledge regarding the subject matter.

The policymakers should find these results useful in developing practical policies and education development programs on the use of ML by identifying the main barriers and the working aspects of the implementation of ML. The knowledge provided by teachers will enable them to assist the learning institutions in designing teaching programs that will enhance technological proficiencies and present new technology solutions. It examines the ethical and operational concerns of ML to help support mechanisms that promote ethical behavior and the sustainability of instructional developments. This is a valuable work because it suggests that teachers are the best individuals to apply in effecting a change in the current educational practices by incorporating technology. This strategy of ML can be used to enhance the learning opportunities and learning outcomes of individuals. The results of this study are a necessity that must be considered in the context of the digital revolution of the education sector and the realization that the revolution is not limited but, rather, must not be an impediment to the educational process.

Following previous scholarly research and research questions, the study derives various hypotheses about the transformation in the views of teachers related to the application of technology in OBE. It is observed that teachers' ease of integration and usefulness perception of ML tools relate positively to their adoption of said technologies. Technical problem level, privacy concerns, and training deficiency could all be taken as negative influencing factors on teacher acceptance of ML tools.

Algorithmic bias and algorithm transparency are ethical issues that also influence educators' ability to adopt these approaches in instructional settings.

The research theoretical basis was developed through the integration of TAM with CLT. According to Davis (1989), TAM indicates that teachers adopt technology when they believe that the technology is useful and convenient. These aspects create the need to consider the benefits that the educational teachers can gain through the use of ML tools in their education and evaluation. The reason is that it is a CLT that Vygotsky (1978) has developed and emphasizes learning as an interaction and scaffold, along with teachers as agents of facilitation in terms of constructing knowledge. Such theoretical frameworks offer a broad methodology to the study of teacher perception as they cover the technical and pedagogical nature of OBE ML adoption. The study project relates the changing technological trends to the educational practice by exploring the views of teaching professionals concerning the use of ML in the OBE settings, which are associated with pragmatic and theoretical advantages to the learning settings.

2 Literature Review

The recent development of Machine Learning (ML) technology brings new opportunities for educational practice transformation through customized teaching and forecast analysis enhancement (Kaddoura & Al Husseiny, 2023). A number of scholarly investigations have established that ML enables personalized educational design according to student profiles, which leads to achievements and better student involvement (Jia, 2025). Research is lacking regarding what educators think about ML tools, their preparedness, and the challenges in applying them to classroom work.

Research shows that ML demonstrates technical efficiency and analytical operations in educational settings (Isaeva et al., 2025). ML algorithms evaluate large assessments and classroom data collections to create adaptive learning system insights (Jose & Vendula, 2024). These systems use performance data to vary instructional materials, which enables the widespread delivery of personalized education. Educational institutions utilize the student outcome prediction capabilities of ML for assessing at-risk students, which enables them to trigger prompt intervention methods that result in enhanced academic achievement.

According to existing academic research, the successful implementation of ML technology in education requires more than just technical capabilities. The utilization of ML technologies in education relies heavily on teachers because they constitute the main conductors of instruction. According to Isabel et al. (2024), the acceptance of ML technologies by educators for classroom practice transformation depends on multiple mappings across practical and perceptual themes. The teaching profession has several obstacles for instructors, including insufficient technical abilities, reluctance to modify conventional teaching practices, and apprehension regarding data security and moral principles (Janja, 2020).

Multiple factors explain why teachers have doubts about new practices. Educators lack sufficient technical preparation for modern ML tools because of inadequate training material targeting their expertise needs. Put simply, insufficient professional growth makes it difficult for teachers to incorporate these sophisticated systems into their educational workflow. Numerous technical challenges prevent educators from integrating ML tools because they need to make them interoperable with Learning Management Systems as well as other educational platforms. The unprepared nature of teachers about these technologies creates stress levels that prevent them from properly utilizing them. Educational institutions continue to face widespread opposition to adopting new educational strategies that stem from their traditional approaches to teaching. Educational practitioners who follow conventional teaching methods find it challenging to trust new ML systems as they rapidly evolve. Data privacy issues, together with concerns about proper student data ethical use, have grown in importance because of today's major digital security and privacy rights concerns (Eden et al., 2024).

Researchers apply the Technology Acceptance Model (TAM) as their main tool for studying the factors that lead teachers to adopt ML tools. Science education research using TAM asserts, through Davis's (1989) proposal, that technology adoption results from perceived ease of use and usefulness measurements. Constructs matter a great deal when reviewing educational spaces. Microlearning tools gain acceptance from teachers when they believe that these technologies produce better learning results and embody smooth usability features that do not create organizational complexities (Buckingham Shum & Luckin, 2019). Multiple studies applying TAM confirm that teachers base their decision to utilize technology in their classrooms specifically on their evaluations of the usefulness, alongside the ease of use of these systems. System functionality and user experience enhancements create major opportunities to boost the educational deployment of these tools (Ali, 2024; Kour et al., 2021; Isabel et al., 2024).

Furthermore, ML can also be evaluated for its educational value based on the evaluation from the CLT, together with the Technology Acceptance Model (TAM). According to Vygotsky's theory in 1978, CLT is based on teacher mediation and illustrates the requirement of environment and individualized feedback in the process of learning construction. Based on theory, it is known that teaching excellence goes beyond simply delivering information due to the demand of instructors to create learning spaces that support students to interact while building concepts through cooperative tasks and organized assistance. This educational approach combines ML tools to provide immediate, tailored feedback for teachers who wish to offer instruction as per the needs of the students. ML systems evaluate classroom exchanges to offer practical advice on how to change how instructors effectively talk with students to improve participation and understanding. The dynamic feedback manner, which is a feature of ML tools, makes them suitable for use in educational teaching purposes because it matches well with CLT.

The combination of TAM and CLT is useful to educational institutions as it explains the crucial aspects regarding classroom ML adoption. In addition to tool capabilities, this pedagogical and technical integration proves that the reception and usage of ML tools can successfully serve pedagogical purposes with appropriate teacher reception and usage. ML has the potential to revolutionize personalized education together with predictive analytics, but complete effectiveness emerges when educators both accept and master the usage of these technologies.

Extensive research on the technological benefits of algorithm-driven platforms in education has not received equivalent attention regarding the factors that affect teachers' adoption behavior. Research demonstrates that professional development programs need to focus on combining technical improvement training for teachers with approaches to build positive attitudes toward technological innovation. Educators can develop interactive settings for sharing knowledge about ML solutions, which reduces teacher resistance to altering their practices (Sharma, 2022). Research demonstrates the requirement of established ethical frameworks together with robust privacy protections to handle bias-related issues, which represents a substantial implementation barrier in educational ML implementation.

The success of predictive learning systems for educational customization and prediction enhancement requires adopting strategies that develop an understanding of teacher-related implementation barriers alongside their acceptance perspectives. Research using TAM and CLT enables a better understanding of the variables that affect teachers' acceptance and failure to adopt ML tools. Success in implementing ML-based interventions depends on resolving these challenges because they impact both implementation success and educational goals focused on inclusive adaptive learning environments. This investigation seeks to resolve current research deficits by examining teachers' behavior towards ML in Outcome-Based Education using systematic methods that provide both scholarly and applied insights for instructional enhancement.

3 Theoretical Framework

The integration of ML within OBE supports the development of individualized learning programs and offers considerable potential to improve educational outcomes via ongoing, data-informed assessment. However, the integration of these new systems is contingent on teachers realizing and adopting these technological systems. To examine teacher perceptions of ML within OBE contexts, this study relies on an integrated theoretical framework of the TAM (Davis, 1989) and CLT (Vygotsky, 1978). The dual framework allows us to complement insights from the technical acceptance of ML tools with their alignment with effective teaching practices.

3.1 Technology Acceptance Model (TAM)

According to Fred Davis, the Technology Acceptance Model of 1989, explains why individuals opt to use new technology. As a matter of fact, it states that there are two key factors: perceived usefulness and ease of use. The perception of usefulness is that the teachers think that the tool will assist them in their work and the process of improving the learning of the students. Therefore, the ML tools that offer more precise feedback, offer personalized feedback, as well as monitor progress attentively, can be considered highly helpful in an Outcome-Based Education (OBE) environment (Roca & Gagne, 2008). The more the teachers are convinced that such benefits will contribute directly to their objectives, such as helping students achieve, the more they will make efforts to persist in using the tool. The second important reason is the perceived ease of use. Based on this, educators would anticipate an easy, intuitive, and low-demand for extra input or education on technology. Those tools do not demand any technical obstacles to utilize and are easily fitting with lesson planning and grading patterns, which are of great interest to educators who already have full schedules (Kivunja, 2014). Teachers will find an ML platform to be useful if it can integrate into the existing work processes with little admonishment. In conclusion, TAM explains that in the case of a teacher adopting new educational technologies, the expected benefits of using the tool are significant, and the degree of its usability is important; actually, both are essential prerequisites.

3.2 Constructivist Learning Theory (CLT)

Constructivist Learning Theory (CLT) can be defined as a learning theory that focuses on enabling people to learn through the processes of learning. Vygotsky (1978) developed his CLT that accentuated the way learners build the latter as a result of social interactions, as well as through experiential learning. According to this theory, good education does not merely entail imparting some knowledge to students, but rather, it entails the establishment of learning environments that encourage inquiry and collaboration with the students, as well as their active participation. The teacher in a constructivist classroom is considered a facilitator who directs the students through the provision of scaffolding and feedback. Summarizing this with regard to OBE, then CLT demonstrates the potential of the use of ML tools as developmental aids. ML systems can provide instant and personalized feedback to learners with unique paths of learning, based on their personal needs. This skill is well placed in the orientation of constructivists on scaffolded learning, where the teachers act to facilitate student learning by providing deeper insights into the interactive student-centered practices. In this respect, the ML scaffolding will help teachers to assist a wider range of learners by providing a range of individual interventions to the learning profiles of different learners (Vygotsky, 1978). The combination of TAM and CLT provides a composite model where the perceptions of teachers regarding ML tools in the OBE situations are analyzed. In the light of TAM, TAM contributes to this by defining the technological aspects, i.e., how teacher attitude towards adoption of ML tools is determined in the light of perceived usefulness and ease of use (Davis, 1989). In its turn, CLT is based on the theory of ML in terms of the pedagogical facets of the ML that support the practices of reinforcement technology in facilitating cooperative learning, scaffolding, and individual learning experiences (Vygotsky, 1978).

3.3 Integrated TAM–Constructivist Theoretical Framework

With the combination of these points, a complex analysis is conducted, and the technical and pedagogical elements of the machine learning adoption can be considered. To be more specific, based on TAM, the key variables are the opinions of teachers on the ease of use and the ease of use (Usability and User Friendliness) of the tools. As an example, teachers consider the possibility of these tools contributing to the improvement of the performance results and to facilitating the assessment process, but to remain in line with the current method of teaching (Kivunja, 2014). CLT pays attention to tools in an attempt to offer personalized, interactive assistance. The tools capable of providing real-time feedback will help in the formation of adaptive learning environments, which result in a change of their instructional practice, the enhancement of teacher-student interaction, and collaborative learning. This theory combined enables researchers to explore the effect of the combination of technical efficiency and pedagogical effectiveness as factors. The combined approach clarifies that these tool perceptions of teachers have their bases on the combination of both technical qualities, in addition to whether the ML tools are compatible with constructivist teaching pedagogy. Another idea that the paper identifies is the possible obstacles (such as technical complexity and resistance to change) to effective integration in OBE programs (Roca & Gagne, 2008). The synthesized theoretical framework of the proposed study in Fig. 1 demonstrates major pedagogic principles of CLT, including scaffolding and personalized instructions, in addition to the TAM constructs of perceived usefulness and perceived ease of use. This model can be used as a guideline for the study of the views of teachers on the implementation of machine learning tools in the Outcome-Based Education (OBE) setting.

The combination of the TAM and CLT provides a more accurate understanding of how machine learning functions in an educational environment. This is because, instead of being concerned about how a tool performs in regard to classroom routines, one is equally concerned with how it will also perform in terms of technical performance, which shows that what is actually important is the integration of the tool in current practices. It has been able to identify key drivers of adoption that teachers would need to employ with students in order to use ClaroRead and has provided the training and support they would require to do so. According to the results, the technologies could become effective in the result-oriented setting,

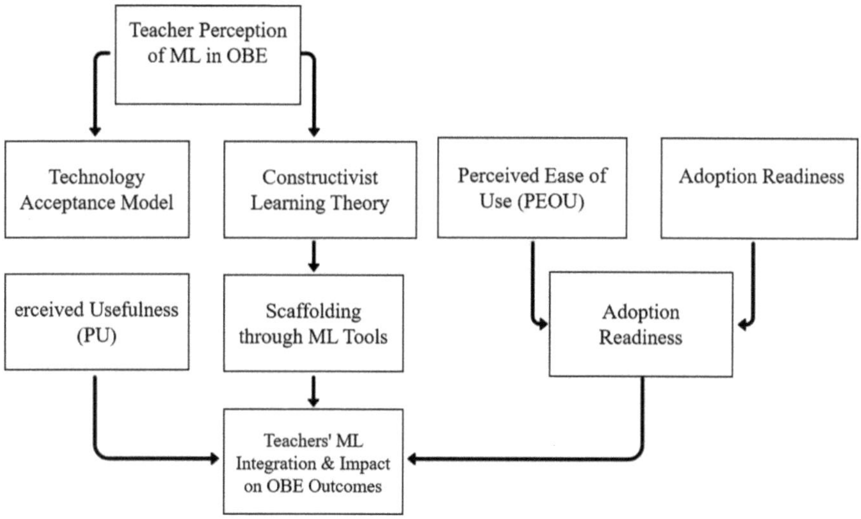

Fig. 1 TAM–constructivist framework of teachers' ML perceptions in OBE

provided that they gain the trust of teachers and that pedagogical objectives are not sacrificed. It is through these revelations that we can inject them into future institutionally-focused data-driven learning programs, enhance professional learning programs, and shape institutional policies to enable us to make a significant and sustainable contribution to classroom practice.

4 Methodology

Research Design To study teacher experiences in machine learning tool integration into Outcome-Based Education environments, the research is qualitative in nature. This qualitative research provides the most appropriate strategy because it enables one to assemble information that indicates deep details on how teachers experience environmental influences when perceiving (Creswell & Poth, 2018). In this research, I perform a detailed qualitative analysis to reveal important factors, barriers, drivers, and enablers of the implementation of ML technologies that quantitative methods alone cannot reveal (Bringle et al., 2023).

4.1 Data Collection Methods

To learn more about teachers' perceptions of these application tools within OBE, the data were collected through combining semi-structured interviews, focus group discussion, the teachers' classroom observations, and their documents.

Semi-structured interviews are the main method for data collection because they enable probing into the participant's experience while still having characteristics of consistency around particular themes (Borucki, 2017). First, participants were given a task to tell their narratives about the platforms, describe what makes it good and what makes it bad, and recommend ways to do ML in action better. Focus group discussions complement the interviews, and during these discussions, there was a focus on talking to and sharing concerns and perspectives on the integration process (Byrne, 2021). Classroom observations for the purpose of understanding the relationship between reflection section reports and observed actual uses of ML tools in the classroom were made to observe the use of actual ML tools in classrooms while teaching was ongoing (Busetto et al., 2020). Finally, the researcher reviewed material such as training manuals and policy guidelines to understand what forms of organizational support and barriers exist that influence the adoption of these platforms. When these methods were combined, they provided a triangulated and comprehensive analysis of the factors that influence teacher engagement with algorithmic platforms in OBE environments.

4.2 Sampling Strategy

The study used purposeful and snowball sampling methods to select the research participants (Greenland & Moore, 2021). Purposeful sampling is used to select teachers who are involved in outcome-based education (OBE) and who have been involved in using ML tools for instruction. It includes active participation in OBE programs and familiarity with educational technologies. The aim is to target this approach to ensure that the participants are properly positioned to furnish more specifics, as well as relevant responses to the research questions (Leighton et al., 2021). Snowball sampling will be used to further increase the variety of viewpoints and to enlarge this sample beyond the facilities associated with this program to diversify the sample of teachers in this study by asking initial participants to recommend other potential participants who also meet the study criteria. Together, these methods generated a rich dataset reflecting diverse integration challenges across Southwest University departments. Table 1 presents the demographic profile of the 18 teachers, including gender, years of teaching experience, subject area, and prior exposure to these platforms.

4.3 Data Analysis

Thematic analysis is a technique that is well-suited to list and interpret patterns within qualitative data and will be used to analyze the collected data. After the data are collected, the research team will start by immersing themselves in the data by reading all interview transcripts, focus group recordings, and note observations

Table 1 Participant demographics (N = 18)

Participant	Gender	Years Teaching	Subject	ML Experience
P1	Male	10	STEM	Beginner (1–2 tools)
P2	Female	4	Humanities	Beginner (1–2 tools)
P3	Male	5	STEM	None
P4	Male	3	STEM	Beginner (1–2 tools)
P5	Female	6	STEM	Beginner (1–2 tools)
P6	Male	12	STEM	None
P7	Female	14	Languages	Intermediate (3+ tools)
P8	Male	9	STEM	Beginner (1–2 tools)
P9	Female	11	STEM	Beginner (1–2 tools)
P10	Female	7	Humanities	Intermediate (3+ tools)
P11	Female	6	Languages	Intermediate (3+ tools)
P12	Female	13	Humanities	Intermediate (3+ tools)
P13	Female	9	Languages	Intermediate (3+ tools)
P14	Female	8	STEM	Beginner (1–2 tools)
P15	Male	7	STEM	None
P16	Female	2	Humanities	Intermediate (3+ tools)
P17	Female	15	Humanities	Beginner (1–2 tools)
P18	Male	8	STEM	None

Note: ML Experience levels are defined by the number of different ML tools participants have used (Author's primary data collection, 2025)

(Braun & Clarke, 2020). This will help in the identification of preliminary codes that will be systematically organized using NVivo software. First, the coding process will take place in the analysis of the text, assigning codes to various text segments, which will relate to certain key themes such as ease of use, perceived usefulness, ethical concerns, and integration challenges (Braun & Clarke, 2022). The initial codes were refined iteratively and grouped into wider patterns that reflected the shared experiences of the participating teachers. In the last stage, these themes are put into the context of the Constructs in the Technological Acceptance Model (Davis, 1989) and CLT (Vygotsky, 1978), which provides a complete view of how pedagogical and technological factors impact ML adoption within OBE environments. Lastly, the systematic use of NVivo will not only guarantee a rigorous analysis of the data, but will also increase the reliability and validity of the study's findings, which are transparent and replicable.

4.4 *Ethical Considerations*

This study was highly ethical, and its rigors guided every step of the research design and implementation (Ross, 2020). Before any data were collected, all participants received clear full information about the purpose of the study, the fact that their involvement would be voluntary, and that they could withdraw at any time without penalty. Prior to the interviews, focus group discussions, or classroom observations, all participants provided informed consent. To ensure the confidentiality of participants, their identities were anonymized through the use of pseudonyms, and all personally identifiable information was removed from transcripts and other research outputs. Furthermore, the research team secured all electronic data on password-protected devices and locked storage (only accessible by the research team) for physical documents. Therefore, these measures were created with the goal of ensuring that the data retains validity and safeguards participants' privacy. The research protocol was reviewed and approved by the IRB of Southwest University in June 2024 (Reference No. SWU-IRB-2024-06-01). Thus, all admitted procedures adhered to the agreed national and institutional standards on what should be done when conducting ethical research with human subjects, as well as how to handle sensitive information. To construct a research environment that was trustworthy and respectful and fostered such a comfortable environment that participants could share honest insights on their experiences with ML tools in the outcome-based education context, research must be subject to these ethical safeguards.

5 Findings and Analysis

This section presents a thematic analysis of qualitative data from 18 educators engaged in outcome-based education (OBE) at Southwest University, Chongqing, and reports the principal findings. Participants varied in their exposure to and use of machine-learning (ML) tools and represented a range of disciplines. Five key themes were identified: (1) perceived usefulness of ML tools, (2) ease of use and integration challenges, (3) personalized learning and student engagement, (4) ethical implications, and (5) professional development needs. Illustrative excerpts from participants support each theme, and Table 2 summarizes the themes, corresponding sub-themes, and the number of illustrative quotes.

Table 2 Themes, sub-themes, and quote frequencies

Theme	Sub-Themes	Frequency of Quotes(*f*)
Perceived usefulness of ML tools	• real-time analytics• deeper behavioral insights•early identification of struggling students	11
Ease of use & integration challenges	• complex interfaces• multi-step workflows• lack of formal training	8
Personalized Learning & Student Engagement	• adaptive difficulty• differentiated instruction• instant feedback	14
Ethical implications	• data privacy concerns• algorithmic opacity• lack of transparency	9
Professional development needs	• hands-on, context-driven training• ongoing support• pedagogical alignment	17

5.1 Perceived Usefulness of ML Tools

One major theme pervading the dataset was the perceived usefulness of ML tools in improving the efficacy of teaching and learning outcomes. The use of ML methods, according to teachers, helped in monitoring performance in real time, performing targeted interventions, and efficient assessment, which are core principles of Outcome-Based Education.

> When I see ML analytics highlighting exactly where my students are having difficulty, I can immediately adjust my lesson plan. It makes my teaching more responsive. **(P7)**

> ML tools help me go beyond traditional grading. They offer insights into student behaviors and engagement levels that I could never track manually. **(P12)**

> The biggest advantage is that I no longer have to wait until the end of the term to find out who's struggling. ML keeps me informed throughout the course. **(P14)**

These insights align with the TAM (Davis, 1989), emphasizing that perceived usefulness is a key determinant of technology acceptance. For the participants, ML tools were not merely technological add-ons but critical instruments that empowered instructional decision-making and student-centered practices.

5.2 Ease of Use and Integration Challenges

Despite acknowledging the benefits of ML tools, participants expressed considerable frustration regarding their usability and the difficulty of integrating them into existing instructional workflows. Usability issues range from complex interfaces to a lack of technical support.

> The system is too complicated for teachers without a strong technical background. We need something more intuitive. **(P 4)**

> There are too many steps to access simple functions. I often give up using the tool mid-class because it interrupts the flow of teaching. **(P 8)**

> We received no formal training. I learned to use the ML software on my own, which took too much time away from actual teaching. **(P 9)**

These observations indicate that although TAM suggests perceived ease of use, until these tools are simple and well supported, they will not be adopted, even if they have all other features (Roca & Gagné, 2008; Kivunja, 2014). This implies that tools used for technology integration need to be designed better, and there is a dire need for a localized support system to facilitate technology integration in educational settings.

5.3 Personalized Learning and Student Engagement

One of the most promising aspects of ML highlighted by teachers is its capacity to support personalized learning experiences and enhance student engagement through adaptive assessments and tailored content delivery.

> One of my students who barely spoke in class became highly active after I used an ML tool to adjust content difficulty based on his progress. **(P1)**

> ML has allowed me to differentiate instruction without having to manually prepare separate lesson plans. It's a huge time-saver and more effective. **(P6)**

> The engagement levels are noticeably higher when students get instant feedback. ML tools give them that feedback in real time, which motivates them. **(P10)**

These perspectives are strongly aligned with CLT (Vygotsky, 1978), which emphasizes scaffolding, individualized learning, and active student involvement. The ability of ML tools to provide real-time feedback and adapt content to student needs underscores their pedagogical alignment with learner-centered approaches in OBE.

5.4 Ethical Implications

Ethical concerns related to data privacy, algorithmic transparency, and student surveillance have been repeatedly raised. Many expressed discomfort with the level of data collection required and uncertainty about how the data was being used.

> I'm not entirely comfortable with how much student data is being collected. I worry about where it's stored and who can access it. **(P 7)**

> The algorithms are black boxes. I don't know how they reach certain conclusions, and that makes me hesitant to rely on them. **(P 11)**

> Students have asked me why a system rated them poorly on participation. I didn't have an answer because the logic behind the scoring wasn't clear to me either. **(P 4)**

These concerns mirror broader debates in the field about the ethical deployment of artificial intelligence in education (Greller & Drachsler, 2012). They highlighted the urgent need for transparent algorithmic frameworks and robust data protection policies that safeguard both students and educators in the digital learning environment.

5.5 Professional Development Needs

A recurring theme across all interviews and focus groups was the lack of adequate professional development to support the implementation of ML tools. Teachers emphasized the need for ongoing, practical training that goes beyond one-off workshops.

> We need more than just a demo session. Real training should involve hands-on experience with case examples from our own classrooms. **(P 3)**

> Even if the tool is good, if teachers don't know how to use it effectively, it's a waste. We need support not just at the start but throughout the semester. **(P 6)**

> Professional development needs to be aligned with pedagogical strategies. It's not just about the software, it's about how to integrate it meaningfully into our teaching. **(P 17)**

These comments suggest that professional development is not merely a logistical requirement but also a strategic necessity for the successful integration of ML in OBE environments. Teachers need continuous, context-sensitive support that bridges the gap between technical functionalities and pedagogical applications.

6 Discussion

6.1 Summary of Key Findings

In a qualitative exploration of how machine learning (ML) can be woven into outcome-based education (OBE) frameworks, teachers identified five pivotal themes: the perceived usefulness of these tools; the hurdles of ease of use and seamless integration; the potential to boost personalized learning and student engagement; ethical considerations around data handling and transparency; and the critical need for ongoing professional development. Overall, educators were enthusiastic about ML's capacity to enrich instruction and tailor learning pathways to individual student needs, yet they stressed that realizing these benefits hinges on having intuitive, well-designed systems, readily available training opportunities, and robust ethical guidelines to ensure responsible, transparent implementation.

6.2 Theoretical Implications

It serves to substantively refine existing theoretical frameworks, especially the TAM and CLT. Davis (1989) developed the TAM, by which perceived usefulness and perceived ease of use are considered major determinants in adopting technology. In this study, although these constructs were reaffirmed by the participants, it was extended to further model within the educational context. In particular, teachers emphasized that the compatibility of ML tools with learner-centered instructional goals—referred to as pedagogical compatibility—is an equally important factor influencing adoption decisions. This indicates that traditional TAM models ignore the importance of instructional relevance in education-oriented technology adoption (Roca & Gagné, 2008; Kivunja, 2014).

Moreover, the findings also resonate with CLT that learning is a socially situated process where learners build knowledge actively by experience or interaction (Vygotsky, 1978). The platforms analyzed during the research were characterized as assisting in scaffolding student learning through real-time performance feedback and the possibility of making differentiated instruction according to the unique learning paths of students. These affordances would be perceived as important in offering autonomy that enables students to be more engaged and support the varied needs of learners, which are key principles in constructivist pedagogy. Therefore, ML tools do not go to waste, and they can be an influential factor in a constructivist teaching method, as long as they are handled with wisdom and under the guidance of the instructor.

6.3 Comparison with Existing Research

The current knowledge in the educational technology area is in agreement with, and adds to, the findings of the research in this study. Nevertheless, other past studies (Ali, 2024; Kour et al., 2021; Isabel et al., 2024) have indicated that machine learning opens opportunities in the investigation of learning analytics, adaptive testing, and performance forecasting. Nevertheless, in most of these studies, the technological potential of these methods was considered, and the perceptions and willingness of users (teachers) were overlooked. This research assumes that the investigator is on the ground level; it is based on the educator's experience. Although Ali (2024) has also noted that educators are opposed to ML due to the absence of training and complexity, his research also shows another facet of complexity since the attitude is formed based on perceived incompatibility between ML tools and the pedagogical framework of OBEs. Moreover, contrary to Greller & Song et al. (2024), who emphasize the potential risks of ethical issues in the context of ML, the research shows how these issues can practically hamper the decision-making process of some teachers to the extent that some teachers will discontinue the use of such systems altogether because of a lack of trust in the information or algorithmic black box. These comparisons substantiate the fact that future ML implementations should transcend the objective of technical optimization and take into consideration the socio-pedagogical environment under which they will be carried out seriously.

6.4 Practical Implications for OBE and Machine Learning Integration

Some of the key findings have a number of implications. To begin with, they should have constant context-sensitive and professional development programs. Teachers also indicated that single training programs will not work and are usually not related to their real instructional challenges. These programs ought to be gradual and progressive, whereby the initial stage is ML literacy, and then advanced to creating an instructional design guided by AI-based insights. They should include case-based training and problem-solving collaboration to gain confidence and fluency in application. Second, the issues of integration will be encountered at both technical and curricular levels. Based on our results, many of the AI-driven platforms do not provide a seamless transition with those tools already available to teachers and are not necessarily related to the achievement of OBE outcomes as stipulated in the curriculum and learning program. In that direction, schools should also build up relations with developers to develop AI-powered tools that will conform to the current OBE rubric and make pedagogy evolve in real-time, taking actionable data into account. This would improve the usability of the systems and also save pedagogical coherence. Third, ethical issues ought to be proactively addressed. In order to get the teachers to be willing to feed data into such systems, we need to ensure that data processing is done in a secure way, when, and where they can know why the decisions are taken in a programmed manner. In addition, there must exist a clear governance policy on such systems, such as periodic review of ethics, algorithmic transparency, and regular revision of data security protocols in institutions. Moreover, the teacher training should be integrated with AI ethics modules, where the trainees of the teaching profession should be informed so that they can be in a position to critically analyze such systems without depending on the knowledge they have acquired.

6.5 Implications for Institutional Policy and Resource Allocation

The schools and universities have already reached the stage of policy choice and are supposed to devote themselves to establishing an ecosystem that could permit the incorporation of machine learning. This involves significant investment in a technology backbone in the form of a stable internet connection, interoperable software platform, and safe data storage platforms. In addition, the institutional policies would assist in maintaining professional learning and technical support as equally important constituents of faculty development instead of enhancing accessories. Motivation to utilize technology can also be offered through recognition or appraisal processes as an effort to encourage innovation. One example that is to be taken into account is the framework of teaching evaluation and program accreditation criteria, which is to be integrated with the incorporation of ML. The interdisciplinary collaboration between the education professionals, Information technology IT professionals, and the policy makers to make decisions in the field should be established in a manner that both the technical and the pedagogical needs are taken into consideration. Besides this, the policymakers ought to enable such initiatives by revising the curricular standards to capture the new digital skills in the wake of the new technologies, as well as investing in technology-enhanced education in a sustainable manner.

6.6 *Limitations of the Study*

This study has several limitations, even though it proposes valuable insights. The qualitative design of the study is effective in yielding in-depth perspectives; however, this provinciality, due to its qualitative design, restricts the generalization of findings to settings other than those being studied. This research was limited to Southwest University in Chongqing, China; thus, the possible views of participants might not represent the views of other institutional or cultural settings. Furthermore, most of the data were self-reported, which may be affected by social desirability bias. While triangulated with observation and document analysis, valid claims could be better supported by student perspectives or the examination of usage analytics. The second focus was on OBE teachers, and finally, the study excluded any study that did not report the discretionary budget of the teachers involved in OBE. Examining educators' ML perceptions from the non–OBE frameworks could provide some points of comparative insight into the impact of pedagogical models on technology acceptance.

7 Conclusion and Recommendations

7.1 *Summary of Findings*

The purpose of this study was to investigate the ways that teachers specifically perceive Machine Learning (ML) tools for use within an Outcome-Based Education (OBE) environment, incorporating the concepts of usefulness, ease of use, ethical concerns, and institutional readiness. Interviews, focus groups, observations of classrooms, and document analysis collected data on the transformative potential as well as the significant barriers to ML adoption in education. Despite teachers' knowledge that ML can tailor teaching and enhance assessment, technical difficulties, insufficient training and resources, data privacy issues, and insufficient system support impede its implementation. The results indicate that, to enable ML integration in OBE, interest in sole advancements in technologies is insufficient, but instead integration in OBE requires alignment with pedagogical values, adequate support systems, and institutional commitment.

7.2 Significance of the Study

The study is a contribution to the body of research on educational technology that illuminates the human aspect of the implementation of machine learning in education and learning. Much emphasis is put on teacher-focused approaches since it is emphasized that teachers have to be prepared, their values must be viewed and addressed, and their worries regarding data privacy and ethical concerns will be taken seriously. It also demonstrates the ways in which these tools can be employed to enhance technicalities in education with regard to handling them properly. Moreover, it introduces useful suggestions to institutions and policymakers regarding the ways of facilitating the effective implementation of ML technologies in educational environments.

7.3 Recommendations

The results of the study are used to provide several recommendations that can help ensure that these tools are successfully introduced in OBE environments. Firstly, schools and colleges must establish and operate customized continuing professional development programs, both technical and pedagogical applications of ML. These programs should be made a perpetual activity by keeping the teachers at all times capable of using machine learning tools in classroom. The peer learning communities must be developed to create collaboration and mutual learning among teachers that will minimize resistance and encourage a sense of community in the adoption process. Second, investing in a strong, efficient technical infrastructure has to be done. These institutions should make sure that these platforms are compatible with the systems that are already in place as Learning Management Systems (LMS) and have consistent access to the Internet and real-time IT services. These are technical requirements, though, which are not optional since, without them, it is impossible to use the technology in the classroom. Also, such tools need to comply with the heuristics of OBE, which include competency-based learning and transparent assessment, among others, and need to be data-driven to offer feedback. Building a partnership between institutions and software developers on such platforms will assist in ensuring that the ML platforms will incorporate the interests of teaching and learning, and not be a peripheral feature. At the policy level, there is an urgent need to provide support for the creation of formal governance structures of ML. The policies are supposed to address ethical issues like privacy and transparency of data, and give some mechanism of accountability in the event of algorithmic bias or data misuse. Lastly, it is important to have sustainable funding to ensure the initiative has long-term sustainability. The allocation of resources must have been made to purchase the technology, to continuously train it, support it, and maintain the infrastructure. This type of strategic investment is to be made on long-range educational goals, along with equality and inclusion principles.

7.4 Future Research Directions

The further research should include longitudinal studies of the change in the attitudes and practices of teachers with increasing the use of machine learning in the process of teaching results as well as mixed methods based studies involving the qualitative studies of the problem with quantitative measures to get a better idea of the role of interactions between teacher perceptions, student results, and organizational conditions in the development of complex problem-solving. The cross-cultural comparison of diverse settings can be used to reflect the adoption and adaptation of machine learning. The outcome measure of professional development could be evaluated on a long-term basis and could be used to explain the impact on the confidence, competence, and continued use of technology by the educators in order to inform scalable capacity-building programs.

References

Ali, Q. I. (2024). Towards more effective summative assessment in OBE: A new framework integrating direct measurements and technology. *Discover Education, 3*(1). https://doi.org/10.1007/s44217-024-00208-5

Anis, M. (2024). Teacher professional development in the digital age: Addressing the evolving needs post-COVID-19. *International Journal for Multidisciplinary Research, 6*(1). https://doi.org/10.36948/ijfmr.2024.v06i01.12386

Arif, M., Irfan, S., Shaoan, M. M. R., & Ali, W. (2025a). AI and machine learning in Islamic guidance: Opportunities, ethical considerations, and future directions. *Guidance: Jurnal Bimbingan Dan Konseling, 21*(2), 429–451. https://doi.org/10.34005/guidance.v21i02.4449

Arif, M., Ismail, A., Aqib, M., Shaoan, M. R., Ali, W., & Okafor, M. U. (2025b). The impact of AI-driven tools on foreign students' Chinese language acquisition: A case study at Southwest University. *Language, Technology, and Social Media, 3*(1), 150–168. https://doi.org/10.70211/ltsm.v3i1.165

Borucki, I. (2017). A visual data collection method: German local parties and associations. *Connect (London), 37*(1–2), 45–52. https://doi.org/10.21307/connections-2017-003

Braun, V., & Clarke, V. (2020). One size fits all? What counts as quality practice in (reflexive) thematic analysis? *Qualitative Research in Psychology, 18*(3), 328–352. https://doi.org/10.1080/14780887.2020.1769238

Braun, V., & Clarke, V. (2022). Conceptual and design thinking for thematic analysis. *Qualitative Psychology, 9*(1), 3–26. https://doi.org/10.1037/qup0000196

Bringle, R. G., Hatcher, J. A., & Jones, S. G. (2023). *International service learning*. Taylor & Francis. https://doi.org/10.4324/9781003445371

Buckingham Shum, S. J., & Luckin, R. (2019). Learning analytics and AI: Politics, pedagogy and practices. *British Journal of Educational Technology, 50*(6), 2785–2793. https://doi.org/10.1111/bjet.12880
Busetto, L., Wick, W., & Gumbinger, C. (2020). How to use and assess qualitative research methods. *Neurological Research and Practice, 2*(1), 1–10. https://doi.org/10.1186/s42466-020-00059-z
Byrne, M. (2021). Interviewing as a data collection method. *AORN Journal, 74*(2), 233–235. https://doi.org/10.1016/S0001-2092(06)61533-0
Creswell, J. W., & Poth, C. N. (2018). *Qualitative inquiry and research design: Choosing among five approaches* (4th ed.). SAGE.
Dahlin, E. (2021). Email interviews: A guide to research design and implementation. *International Journal of Qualitative Methods, 20*(3). https://doi.org/10.1177/16094069211025453
Davis, F. D. (1989). Perceived usefulness, perceived ease of use, and user acceptance of information technology. *MIS Quarterly, 13*(3), 319–340. https://doi.org/10.2307/249008
Dwivedi, Y. K. (2022). Metaverse beyond the hype: Multidisciplinary perspectives on emerging challenges, opportunities, and agenda for research, practice, and policy. *International Journal of Information Management, 66*, 102542. https://doi.org/10.1016/j.ijinfomgt.2022.102542
Eden, O., Emitope, A., Ayeni, O., & Chisom, N. (2024). Inclusive curriculum design: Meeting the diverse needs of students for social improvement. *International Journal of Frontiers in Science and Technology Research, 6*(2), 050–059. https://doi.org/10.53294/ijfstr.2024.6.2.0037
Greenland, S. J., & Moore, C. (2021). Large qualitative sample and thematic analysis to redefine student dropout and retention strategy in open online education. *British Journal of Educational Technology, 53*(3). https://doi.org/10.1111/bjet.13173
Greller, W., & Drachsler, H. (2012). Translating learning into numbers: A generic framework for learning analytics. *Educational Technology & Society, 15*(3), 42–57.
Isabel, E., Bergey, P., & Smith, B. (2024). Blockchain technology for supply chain provenance: Increasing supply chain efficiency and consumer trust. *Supply Chain Management, 29*(4). https://doi.org/10.1108/SCM-08-2023-0383
Isaeva, R., Karasartova, N., Dznunusnalieva, K., Mirzoeva, K., & Mokliuk, M. (2025). Enhancing learning effectiveness through adaptive learning platforms and emerging computer technologies in education. *Jurnal Ilmiah Ilmu Terapan Universitas Jambi, 9*(1), 144–160. https://doi.org/10.22437/jiituj.v9i1.37967
Janja, K. (2020). The future of value in digitalised higher education: Why data privacy should not be our biggest concern. *Higher Education, 2*(2). https://doi.org/10.1007/s10734-020-00639-7
Jia, R. (2025). Practical application and industry relevance: The role of off-campus training in student skill development. *Education and Information Technologies, 2*(2). https://doi.org/10.1007/s10639-025-13408-9
Jose, B.-M., & Vendula, K. (2024). Chat GPT for language learning: Assessing teacher candidates' skills and perceptions using the technology acceptance model (TAM). *Innovation in Language Learning and Teaching, 2*(2), 1–16. https://doi.org/10.1080/17501229.2024.2435900
Kaddoura, S., & Al Husseiny, F. (2023). The rising trend of metaverse in education: Challenges, opportunities, and ethical considerations. *PeerJ computer Science, 9*, e 1252. https://doi.org/10.7717/peerj-cs.1252
Kivunja, C. (2014). Do you want your students to be job-ready with 21st century skills? New pedagogies for new workplaces. *International Journal of Higher Education, 3*(3), 81–91. https://doi.org/10.5430/ijhe.v3n3p81
Kour, S., Kumar, R., & Gupta, M. (2021). Analysis of student performance using machine learning algorithms. In *2021, the Third International Conference on Inventive Research in Computing Applications (ICIRCA)*. https://doi.org/10.1109/ICIRCA51532.2021.9544935.
Leighton, K., Kardong-Edgren, S., Schneidereith, T., & Foisy-Doll, C. (2021). Using social media and snowball sampling as an alternative recruitment strategy for research. *Clinical Simulation in Nursing, 55*, 37–42. https://doi.org/10.1016/j.ecns.2021.03.006

Maghsudi, S., Lan, A., Xu, J., & van der Schaar, M. (2021). Personalized education in the artificial intelligence era: What to expect next. *IEEE Signal Processing Magazine, 38*(3), 37–50. https://doi.org/10.1109/MSP.2021.3055032

Ming, J., & Li, K. C. (2025). Technology-enabled institutional readiness for agile-blended learning: A framework for educational innovation. *SN Computer Science, 6*(3). https://doi.org/10.1007/s42979-025-03822-8

Pusporini, W., & Nurdiyanto, H. (2024). Utilization of artificial intelligence in outcome-based curriculum evaluation and development. *Journal of Research in Social Science and Humanities, 4*(2), 132–134. https://doi.org/10.47679/jrssh.v4i1.109

Rashid, A. B., & Kausik, A. K. (2024). AI revolutionizing industries worldwide: A comprehensive overview of its diverse applications. *Hybrid Advances, 7*(100277), 100277–100277. https://doi.org/10.1016/j.hybadv.2024.100277

Roca, J. C., & Gagné, M. (2008). Understanding e-learning continuance intention in the workplace: A self-determination theory perspective. *Computers in Human Behavior, 24*(4), 1585–1604. https://doi.org/10.1016/j.chb.2007.03.009

Ross, T. W. (2020). Ethical issues of educational information technologies. *Educational Considerations, 23*(2). https://doi.org/10.4148/0146-9282.1438

Sharma, A. (2022). Teacher professional development: Enhancing classroom instruction and student achievement. *Global International Research Thoughts, 10*(1), 10–13.

Sivarajah, U., Kamal, M. M., Irani, Z., & Weerakkody, V. (2021). Critical analysis of big data challenges and analytical methods. *Journal of Business Research, 70*(1), 263–286. https://www.sciencedirect.com/science/article/pii/S014829631630488X

Song, C., Shin, S.-Y., & Shin, K.-S. (2024). Implementing the dynamic feedback-driven learning optimization framework: A machine learning approach to personalize educational pathways. *Applied Sciences, 14*(2), 916. https://doi.org/10.3390/app14020916

Taylor, R., Masoud, F., Ioannou, A., & Konstantina, S. (2024). Personalized learning in education: A machine learning and simulation approach. *Benchmarking: An International Journal, 2*(2). https://doi.org/10.1108/bij-06-2023-0380

Vygotsky, L. S. (1978). *Mind in society: The development of higher psychological processes*. Harvard University Press.

Zamiri, M., & Esmaeili, A. (2024). Strategies, methods, and supports for developing skills within learning communities: A systematic review of the literature. *Administrative Sciences, 14*(9), 1–33. https://doi.org/10.3390/admsci14090231

Continuous Improvement: Strategies for Success

D. Vetrithangam

Abstract Continuous improvement is key to enhancing both student learning and the overall success of educational institutions. This chapter explores how strategic approaches, technology, and machine learning can drive this improvement. It looks at the core principles of continuous improvement in education, including fostering a growth mindset and adopting innovative teaching practices. The chapter highlights how data-driven decision-making can be used to assess and improve student performance, with a focus on tools that track and analyze progress. One key area of focus is the use of machine learning (ML) algorithms, such as Random Forest, to analyze student data, including attendance, test scores, and previous exam performance. These models help categorize students into different performance levels—such as fast, average, or slow learners—allowing educators to understand better where students need support and how to adjust their teaching methods. By doing so, teachers can personalize learning, improve course design, and boost their effectiveness. The chapter also addresses how integrating technology and engaging faculty arc critical for successful continuous improvement. While there are challenges in implementing machine learning in education, this chapter demonstrates how these tools can help identify student strengths and weaknesses, fostering a culture of improvement that benefits both students and educational institutions.

Keywords Continuous improvement · Machine learning · Random Forest · Student's performance

D. Vetrithangam (✉)
Department of Computer Science & Engineering, Chandigarh University, Mohali, Punjab, India

D. Vetrithangam et al. (eds.), *Transforming Outcome-Based Education with Machine Learning*, https://doi.org/10.1007/978-981-95-4925-2_4

1 Introduction

1.1 Defining Continuous Improvement in Education

Continuous Improvement in education refers to an ongoing, systematic approach to enhancing teaching methods, student learning outcomes, and institutional performance. It involves regularly analyzing data, identifying gaps, and making small, continuous changes to improve efficiency and effectiveness (Park et al., 2013; Temponi, 2005). This process uses feedback loops, performance metrics, and technology-driven insights to refine teaching strategies, curriculum design, and student engagement. In modern education, Machine Learning (ML) and Artificial Intelligence (AI) play a crucial role by analyzing student performance data, predicting learning needs, and automating personalized feedback to enhance the quality of education continually.

1.2 Importance of Continuous Improvement for Student Learning &Institutional Success

Similar to education, continuous improvement involves making incremental changes over time. Therefore, students have different learning styles, and some may struggle with specific subjects. It means that teaching methods are continually improved to meet the needs of students. Now, how can any teacher find out what students struggle with in real time? One way is by analyzing the performance of our students through routine assessment and feedback. For instance, if a school identifies that many students are struggling in mathematics, teachers can evaluate test results and the feedback given by pupils who take one of the tests to determine which subjects are the most challenging. How should educators address these challenges once they are identified? By doing so, we can introduce visual learning tools, interactive exercises, and real-world examples to enhance their understanding and provide a more engaging experience of the lessons. However, only relying on traditional assessments will not suffice. If technology can be deployed to help teachers support better learning, then how can that be done? They can utilize machine learning tools to identify patterns in students' performance and suggest personalized learning materials that best cater to their individual strengths and weaknesses. Specific support facilitates quicker improvement and more confidence in learning abilities among students.

Therefore, teachers need to continually cultivate new approaches and refine their teaching methods to ensure students are interested and engaged. For instance, if a university professor notices that students tend to lose interest after 20 min when viewing lecture videos, they can utilize AI-based tools to monitor students' engagement levels. This tool, among other things, shows when students stop paying attention. With this, the professor can divide long lectures into smaller videos that incorporate quizzes, making them more manageable and accessible. Yet, how can teachers confirm that students have grasped the lessons before exams? Quiz answers can be analyzed through AI-powered feedback systems to identify common mistakes made by students, allowing teachers to make instant adjustments to their lesson plans. This enables teachers to improve their teaching based on real-time feedback from students, thereby making learning more interesting and effective.

Improving schools and universities more efficiently through continuous improvement (Hawley, 2006). However, if students are having difficulty accessing open courses and course registration is being delayed at a university, then AI scheduling can help resolve this issue. The expansion of this system handles course registration on a platform that suggests different course options based on student preferences and availability, reducing registration time and making the registration task easier. How can institutions predict what students will need in the future and what resources are necessary to plan? Past enrollment data can be analyzed using machine learning to estimate the number of students who will register for each course. This enables universities to manage faculty assignments and classroom space efficiently (Satterlee, 1996). Continuous improvement in education is also encouraged, with innovation arising from this effort (Parsons, 2017).

Universities can organize regular training programs for teachers, allowing them to test new teaching techniques, digital tools, and AI-based learning platforms, among others. But how can we guarantee to institutions that teachers continually practice and continually improve their teaching skills? Faculty evaluation tools based on AI (Goyal, 2023) can analyze student feedback and learning outcomes to provide an idea about how effective they are as teachers and how they can update their techniques with modern ones. Continuous improvement is also about assessment and feedback. For example, AI grading systems can be used by a college to review students' essays. These systems provide students with instant feedback on grammar, clarity, and structure, allowing them to grow as writers very quickly. How will institutions be able to know that student learning is measured when assessments are administered? AI-driven assessment tools can track learning gaps, identify patterns in student mistakes, and provide insights into how to improve course material. It enables more effective and accurate evaluations, as feedback is provided to students to enhance their performance promptly.

1.3 Role of Technology & Machine Learning in Enhancing Educational Quality

Learning with personalized content, efficiency, and on the go has been made possible due to technology and machine learning (ML). Advancements in education make it easier to improve student outcomes, assist in the professional development of teachers, and help institutions manage their resources more efficiently. However, exactly how does technology help education?

1.3.1 Personalized Learning with Machine Learning

A school utilizes an AI-powered learning system to monitor students' progress (Taylor et al., 2024). The system identifies problem areas if the student is struggling with algebra and suggests that the student first complete simpler lessons before moving on to more complex topics. Additionally, it offers interactive exercises and video tutorials for practicing the concepts. However, this raises the issue of how schools can ensure that each student receives the proper amount of learning support. Adaptive lessons here are based on performance. The system offers additional exercises and step-by-step solutions for students who are not achieving the desired results and are making errors repeatedly.

1.3.2 Enhancing Teacher Effectiveness with AI

Teachers are very important in the education system, but dealing with large classrooms and individual student needs is not always an easy task. It also allows educators to monitor student progress, spot knowledge gaps, and adjust teaching techniques (Pataranutaporn et al., 2021). A high school teacher utilizes an AI system to analyze how students perform on quizzes. The system highlights all topics where students struggle the most. Using this data, the teacher will adjust lesson plans or supplement the material with additional resources for more challenging postings. One of the largest challenges in education is ensuring that students actually understand their lessons before the final exams. Normally, teachers check students' understanding by testing them; however, waiting until the exams may not be possible to assist students who are struggling. Which, in turn, raises an important question: How can teachers be sure that students grasp important topics before the final exams?

However, this problem can be solved using AI-driven feedback tools that instantly analyze quiz responses. The system is triggered to alert the teacher if a large number of students are having difficulty with the topic. As a result, they explain the concept right before the exam to benefit students who struggle, allowing them to learn more effectively and perform better overall.

1.3.3 Smart Assessment and Instant Feedback

Grading by traditional methods is time-consuming, and students may spend days or weeks waiting for feedback. With machine learning, this process is accelerated because students receive instant feedback and can improve immediately (Raghunath et al., 2018). For example, an AI-graded essay evaluation system is introduced in a school. It takes seconds to evaluate grammar, structure, and clarity, and students get suggestions for how to edit and correct.

1.3.4 AI-Powered Course Recommendations and Scheduling

For students and universities, managing both course registration and academic planning is a challenging task. Machine learning also predicts the students' needs and optimizes the class schedule (Amer-Yahia, 2022). Suppose we are in a place where a university experiences class registration problems every semester due to an insufficient number of classes. It utilizes past data on student enrollment and suggests a schedule of courses accordingly to meet the demand of students and the availability of faculty, ensuring an efficient process.

1.3.5 Encouraging Innovation in Teaching and Learning

For continuous improvement in education to be effective, it must be innovative. The new opportunities provided by AI and technology to teachers and students enable them to explore advanced learning methods, making education more interactive and efficient. For example, some schools are utilizing virtual labs enabled by AI, allowing students to complete science experiments and perform other tasks digitally. It enables hands-on learning without the need for expensive laboratory equipment. However, the key problem is to keep teachers up to date with the most modern teaching methods while implementing them. This is facilitated by AI-powered faculty development programs, which provide personalized training sessions. These programs assist educators in adopting innovative teaching strategies and make the technology integration in their classrooms worthwhile and authentic.

2 Foundational Principles of Continuous Improvement in Education

2.1 Growth Mindset & Pedagogical Innovation

This section further elaborates on the essence of continuous education, oriented towards a growth mindset and pedagogical improvement. It offers insights on why teachers and their institutions need to modify and adapt, as well as the way forward, to attain higher learning outcomes.

2.1.1 Growth Mindset

The concept of having a growth mindset can be defined as the idea that a person can enhance their abilities, intelligence, and skills through effort, persistence, and the prospect of learning. Having this mindset enables people to overcome their challenges and continue improving. In comparison, a fixed mindset prevents people from believing they can develop talents or acquire intelligence, instead leading them to think that they are lucky to be good at something or that they are simply not good at anything. For instance, a child who receives low grades in mathematics may claim that they are not good at math. However, when a teacher has a growth mindset, they encourage the student to say, 'I can do better if I practice.' It helps build confidence, motivation, and a desire to dedicate time to learning and growth. The need for a growth mindset is immense because it inspires ongoing learning, encourages individuals to view challenges as opportunities, aids in developing problem-solving skills, and fosters an agile character in every aspect of life.

2.1.2 Pedagogical Innovation

Pedagogical Innovation refers to the adoption of new and effective teaching methods and strategies that enhance learning outcomes, engage students, and respond to the changing needs of the educational landscape. Interactive approaches to education, which include student participation, problem-solving, and real-world applications, have replaced traditional approaches such as one-way lectures or rote memorization. In short, the following are some pedagogical innovations that can enhance students' learning experience.

1. **Flipped Classroom**

By displaying the reversed teaching design, the flipped classroom is essentially a traditional way of teaching, where its form has been flipped or reversed. Instead of being taught something new in class and filling in homework later, students first watch video lectures or study materials at home (Mok, 2014). After this, they join the class while the teacher provides guidance on discussions, problem-solving, and practical activities. This allows students to engage with the material before class, ensuring they are ready for hands-on work. It enables teachers to devote their time to fostering deep understanding and problem-solving, rather than just disseminating information. For example, a physics teacher assigns homework based on a video about Newton's Law. The day after, students conduct experiments in class and discuss their applications in real life. Applying these concepts immediately helps students memorize the concept more effectively in the long run.

2. **Problem-Based Learning (PBL)**

Unlike the problem of memorizing information, the emphasis of Problem-Based Learning (PBL) is on actual problem-solving. Rather than listening to lectures alone, students are put into groups, analyze real-life issues, and find ways to solve them. This improves critical thinking as well as teamwork and research skills. Why is this method effective? This is a real-world problem and enables much greater learning around a more practical and real authentic topic. It teaches students to develop problem-solving skills that they can apply outside the classroom (Kwan, 2009). To illustrate this, let us assume a group of students is placed in a computer science class and assigned a challenge: 'Build an AI for a chatbot in customer support.' Rather than merely studying programming theories, students research, write code, test, and refine their chatbot. For this reason, this hands-on approach to teaching helps to enhance learning and prepare students for real-world applications.

3. **Gamification in Education**

Gamification in education refers to the implementation of game components in educational settings, thereby increasing the level of engagement and interaction during the learning process (Christopoulos and Mystakidis, 2023). Points, badges, leaderboards, and quizzes are among the most common techniques used to motivate students and encourage them to participate actively. Students are more likely to learn and remain engaged when they enjoy the learning process. Furthermore, gamification helps generate a healthy sense of competition among students, aimed at improving themselves. For example, an online quiz about history is designed by a history teacher where students earn points for getting the right answer and receive badges when they complete a topic. This approach to learning is both fun and engaging, which encourages students to participate and stay motivated.

4. **Personalized Learning**

Every student learns differently. Personalized learning means adapting education to each student as an individual/based on his/her needs, learning speed, and interests. Tools powered by AI assist teachers in tracking student progress and provide them with specific exercises tailored to the student's performance (Zhang et al., 2020). In this case, let's say a language learning app learns from a student's progress and refines its lessons accordingly. If the grammar is weak, the app presents another grammar exercise until it is strong enough to move forward. It makes sure that students learn at their own pace and master the concept before moving ahead.

5. **Collaborative Learning**

It aims at students working in teams to solve problems, discuss concepts, and complete tasks. In this way, this method helps in improving communication, teamwork, and leadership skills. It also helps students learn from one another, making it a dynamic learning process. To illustrate this, let's consider a scenario where a biology class is examining the impact of climate change on marine life. Groups of students, each with a different aspect to focus on (marine species, temperature changes, pollution), work together. After completing their research, they compile a group report that combines the results of their individual studies. This collaborative approach fosters teamwork and deeper understanding.

6. **Inquiry-Based Learning**

Inquiry-based learning requires students to ask questions and discover answers rather than receiving information from the teacher (Pedaste et al., 2015). By using this approach, it presents the mind with new ways of creation to develop curiosity, critical thinking, and problem-solving skills. For example, if we consider a real-world case, instead of just picturing photosynthesis, a teacher will ask, 'Why do plants need sunlight?' Before arriving at the right explanation, students conduct experiments, research, and discuss their findings. By this method, learning becomes more enjoyable and arouses students to think like scientists.

2.2 *Data-Driven Decision-Making Supports Continuous Improvement*

2.2.1 Assessment Tools and Attainment Analysis for Academic Performance

The assessment tools are used to measure the student's learning, skill development, and academic performance. Some assessment tools are divided into two types: direct and indirect assessments. Direct assessment tools rely on student performance in either actual academic tasks or simulations of such tasks. Figure 1 shows the framework of the assessment model for evaluating academic performance. This includes exams such as midterms and finals, quizzes to assess understanding, assignments that evaluate research capabilities and analytical skills, projects to

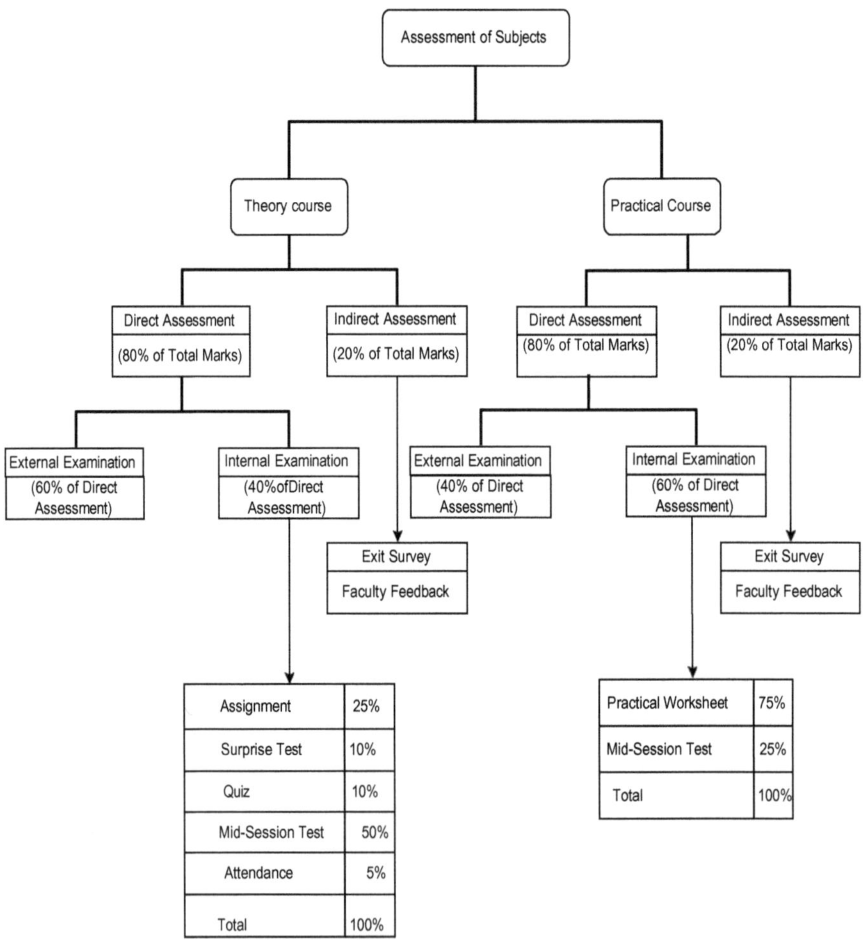

Fig. 1 Framework of the assessment model for academic performance evaluation

apply knowledge in the real world, and lab work or practical exams to test technical skills (Amirtharaj et al., 2022). It collects information through perceptions and feedback, rather than direct testing. Such as surveys and feedback forms, whereby students present their experiences in the learning process, self-assessments that allow students to consider their progress, peer evaluations regarding group work, and course evaluations or alumni surveys to assess the effectiveness of a course. These tools enhance the learning process, as educators can adjust the learning methods according to a student's progress.

They also ensure that courses are effective, are responsible for the continuous improvement of courses, maintain academic standards, and ensure that they meet the accreditation and quality assurance standards required by the institution. Table 1. PO and CO Assessment Tools for Academic Evaluation Purposes. Methods of direct and indirect course appointments are used, which enable the assessment of student learning. Mid-semester tests (twice per semester), semester-end exams, assignments, unit quizzes, and surprise tests (three per course) are included in the direct assessments. Lab courses, summer internships after the fourth and sixth semesters, and project work from the fourth to the eighth semesters are used to evaluate practical skills. The graduate exit survey is one form of indirect assessment that collects feedback on program outcomes from students after they complete the course. These help in a structured way of evaluating students' performance as well as curriculum effectiveness. They are indirect assessment tools used to measure student learning through perceptions, feedback, and self-reflection, not through direct academic learning and performance. Figure 1 illustrates that indirect assessment accounts for 20% of the total assessment.

For example, if students in a programming course struggle to comprehend algorithms, educators can make adjustments to the curriculum or provide additional practice problems in the form of extra work. There are two parts to the Internal Assessment and the External Assessment, which complement the theory examination. Internal Assessment comprises first and second hourly tests, surprise tests, Quizzes, Tutorials, and assignments, which account for 40% i.e., 40 marks. The final examination is in the form of an External Assessment, administered at the end of the semester, and carries a weight of 60 marks (60%) based on all theory courses. Indirect assessment tools evaluate student learning based on the student's perceptions or self-reflection on how they have learned or how they respond to feedback provided by an assessor. It accounts for 20% of the entire assessment process.

Indirect assessment methods enable institutions to evaluate how effectively students are achieving the Program Outcomes (POs) and Program Specific Outcomes (PSOs), as reported by graduates, employers, and parents. For the purpose of understanding academic experience, skill development, and readiness for a professional career, the Graduate Exit Survey is administered to the graduating students who have completed the degree. Every 2 years, the Employer Survey seeks feedback from companies that have hired graduates regarding the graduates' technical skills as well as their ability to solve problems and adapt to the workplace. The Parent Survey is used to gather feedback from parents on their child's academic progress

Table 1 PO and CO assessment tools for academic evaluation

Type of Assessment	Course Assessment and Evaluation Method	Process	Frequency
Direct assessment tools	Mid-semester test	Two midterm tests (MSTs) are conducted per semester to evaluate students' performance. The question paper is framed as per the syllabus, and the questions are aligned with COs and BT levels.	Two in a semester, one after coverage of Unit-1 and the other after coverage of Unit-2
	Semester end examination	At the end of each semester, examinations are conducted for all courses. The questions for this examination cover the entire syllabus of the courses. Question papers are prepared as per the syllabus and aligned with the COs and BT levels.	One at the end of the semester
	Assignments	Assignment question papers are prepared based on the course content and aligned with the course outcomes. It is ensured that the assignment tasks directly address the knowledge, skills, or competencies outlined in the COs.	Total—one assignment per unit
	Technical quizzes	A quiz is conducted per unit according to the assessment pattern to test students' knowledge and reasoning abilities.	One quiz per unit
	Technical training/summer training/ internship	The idea behind training is to provide students with on-the-job training similar to an apprenticeship, more often taken up by university students during the summer in their free time to supplement their formal education and expose them to the world of work extensively with their course, and also to allow them to	Summer training after the fourth and sixth semesters, whose marks are loaded in the fifth and seventh semesters
	Projects	Unique project ideas of faculty and students are shared in a basket for the students to select their project titles. This promotes critical thinking, problem-solving, and the application of learned concepts in practical scenarios.	One per semester (fourth to eighth semester)
	Laboratory courses	According to the rubrics, practical courses are evaluated through a practical worksheet and a viva voce examination.	As per the curriculum

(continued)

Table 1 (continued)

Type of Assessment	Course Assessment and Evaluation Method	Process	Frequency
	Surprise test	The surprise test process begins with the teacher's planning, ensuring secrecy regarding the test date. When administered, students take the surprise test during a class period, completing it within the allotted time. Subsequently, the teacher promptly grades the tests and provides constructive feedback to the students based on their performance, facilitating ongoing learning and improvement.	Three surprise tests are conducted for each course.
Indirect assessment tools	Graduate exit survey	A feedback questionnaire relevant to POs and PSOs is asked in the graduate exit survey form at the end of the program.	Once per program

and career development. Table 2 provides an example of the questionnaire format used in the exit survey form for assessing the attainment of POs and PSOs.

2.2.2 Relation of POs and PSOs with the Questionnaire

The evaluation process is based on a questionnaire consisting of 14 questions that are capable of evaluating both a PO and a PSO. Questions 1 to 12 are aligned to the 12 POs; 13 and 14 are aligned to PSO 1 and PSO 2, respectively. For each question, there are three response options, namely, Extremely Satisfied (3 points), Satisfied (2 points), and Somewhat Satisfied (1 point).

2.2.3 Analysis of Courses Based on Attainment Levels

Student performance is analyzed through attainment analysis, which provides an assessment of how students achieve predetermined learning objectives, such as Program Outcomes (PO) and Course Outcomes (CO). In this way, this process helps identify areas of deficiency in student learning, assesses how well educational goals have been achieved, and provides data-driven evidence of ways to enhance teaching methods and curriculum design. This can be analyzed from the perspective of achieving goals, and depending on the levels, educators can adjust instructional strategies accordingly to ensure continuous improvement (Lavanya and Murthy, 2022).

Table 2 The example questionnaire for stakeholders

Sl. No	Overall, are you satisfied with:
1	Basic knowledge in mathematics, science, engineering, and humanities.
2	Ability to identify, design, analyze, and solve Computer Science & Engineering
3	Design/development of complex engineering problems and their solutions
4	Conduct investigations of complex problems
5	Demonstrate the ability to apply advanced technologies to solve contemporary and new problems.
6	Awareness to apply engineering solutions in global, national, and societal contexts.
7	Understanding professional engineering solutions in societal and environmental contexts
8	Understanding of professional and ethical responsibilities
9	Ability to function as an effective member in multi-disciplinary teams
10	Proficiency in the English language in both communicative and technical forms
11	Demonstrate their ability to independently educate themselves and recognize the significance of continually enhancing their professional knowledge through lifelong learning.
12	Capable of self-education and a clear understanding of the Value of updating their professional knowledge to engage in life-long learning.
13	Rate your ability to apply the knowledge of mathematics, science, and engineering to solve complex problems in the IT field and can design, develop, and provide a solution using modern tools.
14	Rate your ability to apply the knowledge of technology to assess societal, ethical, and management principles to carry out IT projects in diverse environments as a member and leader in a team.

2.2.4 Slow Learner and Fast Learner Identification

For the proper support, educators label students as slow or fast learners. They struggle to grasp the concept, apply the knowledge, and maintain good grades, consistently scoring less than 50% on examinations. Teachers can help students by providing additional classes, employing interactive teaching methods, and monitoring them continuously. Advanced study materials, research opportunities, and leadership roles are offered to fast learners who demonstrate rapid learning and consistently score above 80%. Teachers can identify different learning abilities and create a supportive learning environment that enables all students to succeed.

3 Continuous Improvement Methodologies for Education

Continuous improvement in education refers to an ongoing, data-driven process that enhances teaching strategies, curriculum effectiveness, and student learning outcomes. By systematically analyzing educational performance, institutions can implement targeted interventions to improve academic quality and institutional effectiveness.

3.1 Lean Principles in Academic Institutions

Lean principles in academic institutions focus on maximizing value while minimizing waste. This involves streamlining processes, improving efficiency, and enhancing the quality of education and services. By applying lean techniques, institutions can optimize resources, reduce unnecessary steps, and improve overall student satisfaction and outcomes (Jahan and Doggett, 2015; Doman, 2011). This also helps reduce inefficiencies in teaching and administrative processes, streamlining curriculum design to achieve better learning outcomes.

3.2 Six Sigma in Higher Education

Six Sigma in higher education focuses on improving processes and reducing errors in academic institutions. By using data-driven methods, it aims to enhance student outcomes, streamline administrative tasks, and ensure consistent quality in teaching. The approach emphasizes continuous improvement through defined processes and measurable goals (Antony et al., 2012). This includes identifying and eliminating variability in student performance, as well as conducting data-driven evaluations of teaching methodologies to enhance effectiveness.

3.3 Kaizen for Incremental Educational Improvements

Galbraith is also the author of his book on Kaizen, a Japanese term which means " continuous improvement ", and which focuses on making slow, incremental changes to improve efficiency and quality. Kaizen in education helps institutions adopt better teaching methodologies, curriculum design, student involvement, and overall academic performance through a culture of learning and improvement (Mahon, 1991; Wiid, 2018). Education pursued in the context of kaizen aims at improving learning through minor, continuous improvements. In other words, teachers, students, and administrators collaborate to create a more effective learning environment. Based on student performance data, teaching methods and curriculum are continually improved. The approach is student-centered, which helps one meet the needs of individual learning, and therefore is effective. Unnecessary processes need to be removed. Training the teacher is beneficial as it enables them to improve their skills. Such feedback and reflection on teaching serve as continuous feedback to educators, who then have opportunities to review and improve their strategies, thereby promoting ongoing educational improvement.

3.4 Long-Term Success: (PDCA) Implement Small Continuous Changes

The results from assessment and evaluation are the main indicators for improving the program through the Philosophy of Outcome-Based Education. This results in continuous improvement on Program Educational Objectives (PEOs), Program Outcomes (POs), Program Specific Outcomes (PSOs), Course Outcomes (CO), Student Outcome (SO), curriculum, teaching-learning process, assessment methods, feedback mechanisms, and faculty training in future skills as per the Plan-Do-Check-Act (PDCA) cycle (Moen and Norman, 2006). Departments assess the COs attained at the end of each semester and identify gaps; they then propose action plans (which are subject to change depending on the results of the following semester) for closing the gaps, as shown in Fig. 2. Following this, the following recommendations are put forward for further improvement to the Academic Council.

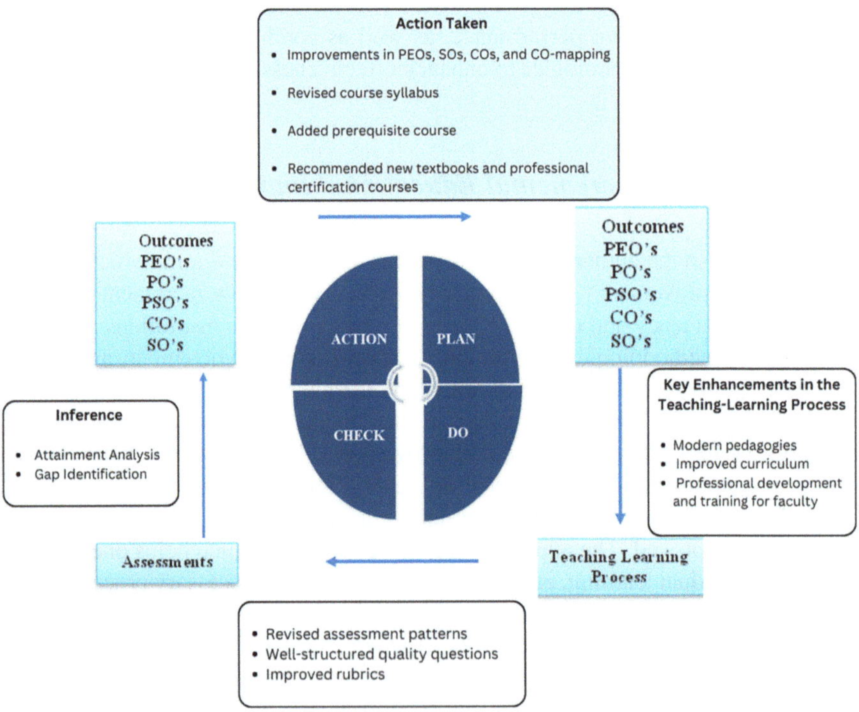

Fig. 2 PDCA Cycle for continuous improvement

3.5 Case Study on PDCA-Driven Improvements in Education: Enhancing Outcomes in Microcontroller and Embedded Systems

A university department offering Microcontroller and Embedded Systems observed varying student performance across different Course Outcomes (COs). To ensure better learning outcomes, the academic committee seeks a detailed analysis. Evaluate student attainment of COs using assessments, identify key learning gaps and challenges, and provide recommendations from the academic council meeting.

For instance, if the course is Microcontroller and Embedded Systems, and there are 140 students for the internal test. Specifically, the test covers all Course Outcomes (CO1—CO5) and has the following attainment values: CO1 = 2.0, CO2 = 1.9, CO3 = 1.7, CO4 = 1.6, and CO5 = 1.8, with an overall attainment value of 1.8. Setting the target at 65% implies, negatively, that the course has not achieved its target. This necessitates that remedial actions be taken for slow, average, and fast learners, depending on the institution's needs.

3.5.1 Gaps, Remedial Measures, and Recommendations to the Academic Council

The course is primarily theoretical, hence students have limited practical experience. It is difficult for them to understand the practicality of the term 'interrupt handling'. The course covers how to configure interrupts, write an interrupt service routine, and manage interrupt priority; however, most students find it difficult to practically implement these concepts. The gap between this theory and the skills a student needs to implement and debug interrupt-driven programs can be partially bridged by hands-on labs using development boards, such as Arduino or PIC microcontrollers.

The following recommendation is also forwarded to the Academic Council to offer this course 23CST-202 'Microcontroller and Embedded Systems' to the students as an experiential course with an option for a mini project included in the assessment methods.

3.5.2 Action Taken for Slow Learners

1. Remedial classes will be conducted for students who fall into the slow learner category.
2. Recommendation of introductory MOOC course through Coursera/ LinkedIn

   a. The Arduino Platform and C Programming

(Approx. 12 hours)
https://www.coursera.org/learn/arduino-platform

b. Microcontroller and Industrial Applications

(Approx. 13 hours)
https://www.coursera.org/learn/microcontroller-and-industrial-applications

c. Learning Arduino: Foundations

(1 hour 6 mins)
https://www.linkedin.com/learning/learning-arduino-foundations

3.5.3 Action Taken for Fast Learners

1. Students will be trained on microcontrollers and embedded systems by working on hands-on projects involving sensor integration and control mechanisms. They will learn to program microcontrollers to interface with hardware components, gaining practical experience in real-world applications like home automation and IoT devices
2. The advanced certification courses will be recommended from Coursera/ LinkedIn

a. Learning Arduino: Interfacing with Hardware

(1 hour 40 mins)
https://www.linkedin.com/learning/learning-arduino-interfacing-with-hardware

b. Learning Arduino: Pulse Width Modulation

(1 hour 16 mins)
https://www.linkedin.com/learning/learning-arduino-pulse-width-modulation

c. Interfacing with the Arduino

(Approx. 11 hours)
https://www.coursera.org/learn/interface-with-arduino

d. Programming for the Internet of Things Project

(Approx. 6 hours)
https://www.coursera.org/learn/internet-of-things-project

3. Students are motivated to participate in competitions like hackathons and hands-on workshops inside and outside the university.

3.6 *Machine Learning for Continuous Improvement*

As Machine Learning continuously evolves and improves, it is driving a transformation in education. Personalization, efficiency, and adaptability to different learning needs are often not the case with traditional teaching and assessment methods. Using Machine Learning, data-driven decision-making is enabled, repetitive tasks are automated, and student learning experiences are enhanced through intelligent tutoring systems, adaptive assessments, and predictive analytics (Gupta et al., 2022). ML analyzes student performance, learning behavior, and engagement patterns, enabling educators to customize instructional strategies, optimize curricula, and improve student outcomes. However, the application of ML does not stop at individualized learning, as it also contributes to making institutional decisions, faculty workload distribution, dropout prediction, and curriculum improvement. Consequently, the analysis of vast educational data enables policy-making and curriculum design informed by evidence, ensuring the ongoing development of effective teaching methodologies and student success strategies.

3.7 *Machine Learning for Personalized Learning and Student Success*

The education system is being transformed by machine learning to make learning more personalized, efficient, and data-driven. Various ML models help in analyzing students' performance, automating grading, and improving teaching strategies. Decision Trees and Random Forest, supervised learning models, predict student success, identify struggling students, and help suggest personalized learning plans (Kaswan et al., 2024). The methods of support vector machines (SVM) are used for automated grading and plagiarism detection, while Naive Bayes is employed for predicting semantic tags in students' feedback. Unsupervised learning models, such as K-Means Clustering, help educators identify which students belong in a particular group based on their learning styles and levels of engagement, allowing them to target their support more effectively. PCA helps identify the primary factors that affect performance. Adaptive learning systems that move with learners, highly engage students, and retain more knowledge rest on the shoulders of Reinforcement Learning (RL)—the AI tutor adjusts the lesson based on student response. Additionally, it assists in optimizing the class schedules and the eLearning resources. Recurrent Neural Networks (RNNs) and Long Short-Term Memory (LSTM) models consider the learning patterns of students to predict their future progress. In particular, tools such as automated handwriting assessment and interactive learning applications rely on CNNs. Several AI chatbots provide instant tutoring support, which is powered by models such as BERT and GPT, both of which are based on the transformer architecture. Additionally, adaptive learning can be achieved with hybrid models, which combine various techniques. For instance, integrating feature selection with Random Forest helps predict dropout risk for students, allowing interventions to be implemented in a timely manner. Khan Academy (Sylvie Beatrice et al., 2024) is a real-world example of ML (AI-powered) being used in the education field to recommend which subjects a student should focus on improving based on their past performance. Just like that, Coursera and Duolingo also use ML to personalize courses and alter their difficulty level so that everyone can learn better.

3.8 ML Algorithms to Find the Learning Patterns for Students

These days, ML is commonly used in the field of education. Meanwhile, as more data about student performance becomes available, ML helps teachers and researchers study how students learn and identify patterns, as well as predict outcomes such as exam scores. In this scenario, by running the tests, ML can provide valuable insights to help improve teaching methods and also offer support to students.

As an example, let's take the case of real-time. We are given a dataset consisting of 70 students, who have attributes such as Student ID, Student Name, and Attendance Percentage, as well as the marks each student received for each question. Q1 to Q8, and to what extent can the machine learning model help classify the students into the performance category (High, Medium, Low), and can it pinpoint the strengths and weaknesses of students in different knowledge levels (K1–K6) and course outcomes (CO1–CO5)? With wisdom, teachers can empower themselves with Machine Learning (ML) algorithms that help them understand how their students perform and learn. With the help of ML, we can analyze data on test scores, attendance, K1 to K6 knowledge levels, and CO1 to CO5 course outcomes to detect patterns that indicate areas where students excel or struggle.

For example, we can utilize algorithms such as decision trees or random forests to classify students as high, medium, or low performers based on their scores. They can also indicate which factors (such as attendance or level of specific knowledge) are influential on performance. It enables teachers to take early steps to support students who are struggling, modify teaching methods, and refine course design. ML can even suggest personalized learning strategies for each student. However, to effectively utilize ML in education also requires high-quality data, accurate student information, and models that teachers can trust and understand.

To study student performance and group them into categories such as Fast Learner, Average Learner, or Slow Learner, we can use the Random Forest algorithm by following these steps.

```
Input:
 Student Dataset D with attributes:
 Attendance, the weightage of marks for questions, Total_score,
and Previous exam_score
 Performance Labels: Fast Learner, Average Learner, Slow Learner

Output:
 Trained a Random Forest model to classify students
 Performance category for each student
 Feature importance for learning insights

Step 1: Data Preprocessing
 For each record in D:
 - Remove or handle missing values
 - Normalize scores (e.g., K1 out of 2, CO1 out of 4)
 - Calculate overall percentage score
 - Assign performance label:
 If score ≥ 80% → Fast Learner
 Else if 50% ≤ score < 80% → Average Learners
 Else → Slow learner

Step 2: Feature Selection
 Select features:
 F = [Attendance, Q1-Q8, K1-K6, CO1-CO5]

Step 3: Split Dataset
 Randomly split D into:
 - Training Set (e.g., 80%)
 - Test Set (e.g., 20%)

Step 4: Build Random Forest Model
 Define parameters:
 N = number of trees in the forest
 M = number of random features to consider at each split

 For i = 1 to N:
 - Draw a bootstrap sample Di from the training set
 - Build a decision tree Ti on Di using:
 - Random selection of M features at each node
 - Best split based on Gini Impurity or Entropy

 Store all trees: Forest = {T1, T2,..., TN}
```

```
Step 5: Classification
 For each student record x in the test set:
 - Predict label from each tree Ti → label_i
 - Final label = majority vote(label_1, label_2,..., label_N)

Step 6: Evaluation
 Compare predicted labels with actual labels
 Compute metrics:
 - Accuracy
 - Precision
 - Recall
 - F1-score

Return:
 Trained a Random Forest model
 Student performance predictions
 Ranked list of essential features
```

Figure 3 illustrates the performance of students across different knowledge levels (from K1 to K6) in terms of the percentage of maximum marks they scored. Out of these, K3, with its lowest score (52.57%), appears to be the most challenging for students to master. Only slightly better results are achieved by K5 (61.14%) and K6 (62.00%), both of which could be improved. On average, the level of understanding, K1, is slightly stronger at 63.21%. In contrast, students performed well in K4 (78.57%) and K2 (78.93%), indicating they are pretty confident in these topics. To understand how this varies, we used the Random Forest algorithm, which can also learn about different factors, such as how often students attend classes, their

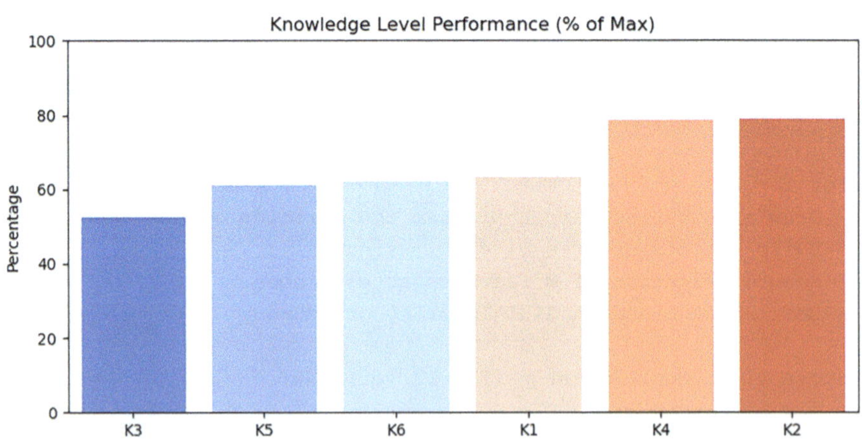

Fig. 3 ML-based analysis of student performance across knowledge levels

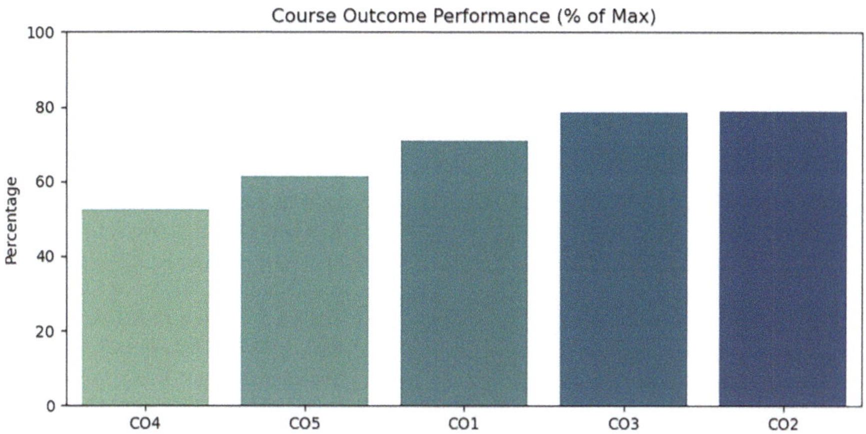

Fig. 4 Course outcome-based student performance analysis with ML techniques

performance on various types of questions, or their internal test results. This enables teachers to gain a better understanding of where students already excel and where they may need additional support. With such insight, educators can create lessons and support strategies that are focused on helping every student improve and succeed.

In Fig. 4, Course Outcome Performance (% of Maximum Marks) indicates the performance of students across different Course Outcomes (CO1 to CO5). With the least performance, CO4 shows a performance of 52.57%, indicating that students find this outcome the most taxing. CO5 has a moderate performance at 61.57% whereas CO1 is slightly high, coming in at 71.07%, meaning that they understand the reading, but at the same time, they still have room for improvement. As shown in Table 3, students achieve the best results with CO3 and CO2 (78.57% and 78.93%) they confirming a good understanding. With the help of the Random Forest algorithm, we can pinpoint the attributes that are responsible for such variations, such as attendance, marks allotted in each question, Total score, and previous test score. This enables educators to create customized learning aids and support mechanisms to address areas of weakness and enhance the overall measure of student achievement.

4 Faculty Engagement & Technology Integration for Continuous Improvement

Continuous improvement in education ensures that both teaching methods and student learning outcomes evolve in response to advancements in technology and pedagogy. Both faculty members and the technological integration process play an essential role, with Artificial Intelligence (AI) providing additional support for their effectiveness. The following description outlines the methods for faculty participation in machine learning (ML) and Artificial Intelligence (AI) to enhance educational support for students.

Table 3 Individual student strengths and weaknesses analysis

Individual Student Strengths and Weaknesses
Student 1 = > Best K: K1, Worst K: K5 | Best CO: CO1, Worst CO: CO5.
Student 2 = > Best K: K1, Worst K: K3 | Best CO: CO3, Worst CO: CO4.
Student 3 = > Best K: K1, Worst K: K3 | Best CO: CO1, Worst CO: CO4.
Student 4 = > Best K: K2, Worst K: K3 | Best CO: CO1, Worst CO: CO4.
Student 5 = > Best K: K2, Worst K: K6 | Best CO: CO2, Worst CO: CO4.
Student 6 = > Best K: K1, Worst K: K3 | Best CO: CO1, Worst CO: CO4.
Student 7 = > Best K: K1, Worst K: K6 | Best CO: CO2, Worst CO: CO4.
Student 8 = > Best K: K2, Worst K: K3 | Best CO: CO2, Worst CO: CO4.
Student 9 = > Best K: K1, Worst K: K5 | Best CO: CO1, Worst CO: CO5.
Student 10 = > Best K: K2, Worst K: K3 | Best CO: CO2, Worst CO: CO4.
Student 11 = > Best K: K2, Worst K: K3 | Best CO: CO2, Worst CO: CO4.
Student 12 = > Best K: K2, Worst K: K3 | Best CO: CO2, Worst CO: CO4.
Student 13 = > Best K: K1, Worst K: K3 | Best CO: CO1, Worst CO: CO4.
Student 14 = > Best K: K1, Worst K: K4 | Best CO: CO1, Worst CO: CO3.
Student 15 = > Best K: K4, Worst K: K5 | Best CO: CO3, Worst CO: CO5.
Student 16 = > Best K: K4, Worst K: K3 | Best CO: CO3, Worst CO: CO4.
Student 17 = > Best K: K1, Worst K: K5 | Best CO: CO1, Worst CO: CO3.
Student 18 = > Best K: K1, Worst K: K3 | Best CO: CO1, Worst CO: CO4.
Student 19 = > Best K: K1, Worst K: K3 | Best CO: CO2, Worst CO: CO4.
Student 20 = > Best K: K3, Worst K: K6 | Best CO: CO2, Worst CO: CO1.
Student 21 = > Best K: K2, Worst K: K3 | Best CO: CO1, Worst CO: CO4.
Student 22 = > Best K: K2, Worst K: K1 | Best CO: CO2, Worst CO: CO4.

(continued)

Table 3 (continued)

Individual Student Strengths and Weaknesses
Student 23 = > Best K: K4, Worst K: K3 I Best CO: CO3, Worst CO: CO4.
Student 24 = > Best K: K4, Worst K: K1 I Best CO: CO3, Worst CO: CO5.
Student 25 = > Best K: K4, Worst K: K1 I Best CO: CO3, Worst CO: CO4.
Student 26 = > Best K: K4, Worst K: K5 I Best CO: CO3, Worst CO: CO2.
Student 27 = > Best K: K1, Worst K: K3 I Best CO: CO5, Worst CO: CO4.
Student 28 = > Best K: K4, Worst K: K1 I Best CO: CO3, Worst CO: CO4.
Student 29 = > Best K: K2, Worst K: K3 I Best CO: CO2, Worst CO: CO4.
Student 30 = > Best K: K1, Worst K: K3 I Best CO: CO3, Worst CO: CO4.
Student 31 = > Best K: K1, Worst K: K2 I Best CO: CO3, Worst CO: CO1.
Student 32 = > Best K: K4, Worst K: K1 I Best CO: CO3, Worst CO: CO4.
Student 33 = > Best K: K1, Worst K: K5 I Best CO: CO1, Worst CO: CO4.
Student 34 = > Best K: K1, Worst K: K3 I Best CO: CO1, Worst CO: CO4.
Student 35 = > Best K: K5, Worst K: K1 I Best CO: CO5, Worst CO: CO1.
Student 36 = > Best K: K1, Worst K: K3 I Best CO: CO2, Worst CO: CO4.
Student 37 = > Best K: K2, Worst K: K1 I Best CO: CO2, Worst CO: CO1.
Student 38 = > Best K: K1, Worst K: K3 I Best CO: CO3, Worst CO: CO4.
Student 39 = > Best K: K1, Worst K: K5 I Best CO: CO3, Worst CO: CO5.
Student 40 = > Best K: K5, Worst K: K1 I Best CO: CO5, Worst CO: CO1.
Student 41 = > Best K: K2, Worst K: K1 I Best CO: CO2, Worst CO: CO1.
Student 42 = > Best K: K1, Worst K: K2 I Best CO: CO3, Worst CO: CO5.
Student 43 = > Best K: K3, Worst K: K5 I Best CO: CO2, Worst CO: CO5.
Student 44 = > Best K: K6, Worst K: K5 I Best CO: CO2, Worst CO: CO1.

(continued)

Table 3 (continued)

Individual Student Strengths and Weaknesses
Student 45 = > Best K: K3, Worst K: K1 I Best CO: CO3, Worst CO: CO1.
Student 46 = > Best K: K2, Worst K: K5 I Best CO: CO2, Worst CO: CO5.
Student 47 = > Best K: K3, Worst K: K1 I Best CO: CO4, Worst CO: CO1.
Student 48 = > Best K: K1, Worst K: K5 I Best CO: CO2, Worst CO: CO4.
Student 49 = > Best K: K1, Worst K: K2 I Best CO: CO3, Worst CO: CO2.
Student 50 = > Best K: K2, Worst K: K1 I Best CO: CO2, Worst CO: CO1.
Student 51 = > Best K: K3, Worst K: K5 I Best CO: CO3, Worst CO: CO5.
Student 52 = > Best K: K1, Worst K: K3 I Best CO: CO1, Worst CO: CO4.
Student 53 = > Best K: K4, Worst K: K3 I Best CO: CO3, Worst CO: CO4.
Student 54 = > Best K: K1, Worst K: K6 I Best CO: CO2, Worst CO: CO3.
Student 55 = > Best K: K1, Worst K: K3 I Best CO: CO1, Worst CO: CO4.
Student 56 = > Best K: K1, Worst K: K4 I Best CO: CO5, Worst CO: CO3.
Student 57 = > Best K: K1, Worst K: K3 I Best CO: CO5, Worst CO: CO4.
Student 58 = > Best K: K2, Worst K: K1 I Best CO: CO2, Worst CO: CO1.
Student 59 = > Best K: K3, Worst K: K5 I Best CO: CO4, Worst CO: CO5.
Student 60 = > Best K: K4, Worst K: K5 I Best CO: CO3, Worst CO: CO5.
Student 61 = > Best K: K1, Worst K: K6 I Best CO: CO1, Worst CO: CO4.
Student 62 = > Best K: K3, Worst K: K6 I Best CO: CO3, Worst CO: CO5.
Student 63 = > Best K: K4, Worst K: K1 I Best CO: CO3, Worst CO: CO1.
Student 64 = > Best K: K1, Worst K: K3 I Best CO: CO2, Worst CO: CO4.
Student 65 = > Best K: K5, Worst K: K3 I Best CO: CO1, Worst CO: CO4.
Student 66 = > Best K: K2, Worst K: K1 I Best CO: CO2, Worst CO: CO1.

(continued)

Table 3 (continued)

Individual Student Strengths and Weaknesses
Student 67 = > Best K: K1, Worst K: K3 I Best CO: CO1, Worst CO: CO4.
Student 68 = > Best K: K4, Worst K: K1 I Best CO: CO3, Worst CO: CO4.
Student 69 = > Best K: K4, Worst K: K5 I Best CO: CO3, Worst CO: CO2.
Student 70 = > Best K: K2, Worst K: K3 I Best CO: CO1, Worst CO: CO3.

4.1 Encouraging Faculty Participation in Continuous Improvement

Faculty members must learn new teaching techniques and innovative approaches to continuously improve education. There are professional development programs and reward systems that can be leveraged to encourage this. Teacher professional development programs provide opportunities for teachers to stay updated with modern teaching strategies and technologies through workshops, online courses, and seminars, where they learn about current teaching trends, effective student engagement strategies, and how to utilize data for improved learning outcomes. One example could be a university that holds monthly training sessions for faculty to teach them how to use AI tools to track students' performance and adapt their teaching based on the data provided. Teachers should also be recognized and rewarded for using creative teaching approaches, as this may encourage them to work even harder. It can be awards, salary bonuses, promotions, or public appreciation. One example might be a college providing an 'award for the best innovative teacher' to those teachers who experiment with new forms of assessment or novel classroom activities, and they would help others in the same way.

4.2 AI for Better Teaching and Learning Support

The teacher began to learn differently, and as a result, how students learn differently started to change due to Machine Learning (ML). Currently, teachers can monitor the students' performance in real-time using AI-based tools. For instance, in online quizzing, the teacher can see which questions most students answer incorrectly and then plan a revision session on those topics. With Smart Learning Management Systems (LMS), students' scores, attendance, and participation can be analyzed automatically, and instant feedback can be provided or suggestions for improvement can be offered. They pull help from students quickly when they make mistakes, and they're staying on track. AI also supports students through smart systems.

Early warning tools are used to identify students who are underperforming or frequently missing classes. For example, if a student's attendance is less than 60% and their marks are also low, the system can issue an alert to the teacher to take appropriate steps beforehand. Furthermore, AI chatbots and virtual assistants are constantly online to answer students' queries, such as deadlines, exam schedules, or subject-related questions, without requiring human tutor assistance. It is also proving to help make the student assessment more accurate and faster. Predictive analytics can identify potential student failure or drop-out and suggest ways to help them as early as possible. Natural Language Processing (NLP) is the backbone of AI grading systems, as they use it to specifically check written answers and assignments, which saves teachers time and provides students with quicker feedback. According to adaptive learning platforms, it can suggest learning materials depending on what a student is comfortable with and where he/she need improvement. In such cases, they can also add new content to the syllabus, adjusting according to the class's learning speed and interest. They help make education more personalized, better, and friendlier for students.

5 Overcoming Challenges in Implementing Continuous Improvement with Machine Learning

Machine Learning (ML) and Artificial Intelligence (AI) can bring significant benefits to education, but they also present some challenges in this regard. The issue is that many teachers may not be confident using AI Tools, due to a lack of technical ability or being accustomed to the traditional way of teaching. For example, a mathematics professor at a college may hesitate to incorporate a predictive analytics tool into their classes because they are unsure about how to interpret or apply the tool's results. Another major issue is ethics, specifically how and why student data is collected and used. If not, the right things are put in place to protect them, schools may worry that AI would make unfair decisions or breach student privacy.

Institutions can conduct training programs for teachers to help them understand the effective use of AI tools. For instance, a university could also host monthly workshops for university faculty to help them understand how to use an AI-powered LMS to measure student performance in real time and give real-time feedback. Furthermore, colleges can establish clear guidelines on how to handle data responsibly and protect the privacy of their students. A good example is the encryption of a data system, with only authorized staff having access to sensitive information. Schools can then utilize AI and ML through continuous faculty skill development and data security to achieve better teaching and student success outcomes.

6 Conclusion & Future Directions

Continuous improvement in education isn't about reaching a final goal—it's an ongoing journey of refining how we teach and how students learn, all while helping institutions grow stronger. In this chapter, we explored how combining traditional educational values with modern tools, such as machine learning—specifically, the Random Forest algorithm—can make this journey more effective. By examining student data closely, including attendance, test scores, and past performance, we identified valuable patterns in learning outcomes. For example, students performed particularly well in CO2 and CO3, while CO4 showed lower performance in the dataset we analyzed, indicating areas that require more focused support. These insights enable teachers to design more effective and personalized learning experiences, making informed adjustments to their course plans. Of course, real progress also depends on active faculty involvement and smart use of technology. While there are challenges, embracing data-driven innovation leads to a more flexible and supportive learning environment—one that truly benefits both students and institutions.

This chapter only explains how machine learning can enhance education, but there are many other avenues worth exploring. However, future work may further expand the data to include more student behaviors and learning environments in addition to what is seen here, such as online activity, project performance, or peer reviews. Another path to explore could be to use deep learning models as an additional tool to examine the problem, in conjunction with standard techniques, such as Random Forest, for greater detail. Another exciting area is the ability to build real-time dashboards for educators, allowing them to visualize their students' progress and make adjustments on the fly in their teaching. It would also be interesting to study whether these data-driven approaches have any impact on long-term student outcomes, such as retention, skill development, and employability. Finally, involving students in the feedback process with appropriate insights for them will enhance their self-awareness and self-motivation. Thus, this would facilitate the development of a more collaborative and responsive education system.

References

Amer-Yahia, S. (2022). Towards AI-powered data-driven education. *Proceedings of the VLDB Endowment, 15*(12), 3798–3806.

Amirtharaj, S., Chandrasekaran, G., Thirumoorthy, K., & Muneeswaran, K. (2022). A systematic approach for assessment of attainment in outcome-based education. *Higher Education for the Future, 9*(1), 8–29.

Antony, J., Krishan, N., Cullen, D., & Kumar, M. (2012). Lean six sigma for higher education institutions (HEIs): Challenges, barriers, success factors, tools/techniques. *International Journal of Productivity and Performance Management, 61*(8), 940–948.

Christopoulos, A., & Mystakidis, S. (2023). Gamification in education. *Encyclopedia, 3*(4), 1223–1243.

Doman, M. S. (2011). A new lean paradigm in higher education: A case study. *Quality Assurance in Education, 19*(3), 248–262.
Goyal, A. (2023). Driving continuous improvement in engineering projects with AI-enhanced agile testing and machine learning. *International Journal of Advanced Research in Science, Communication and Technology, 3*(3), 1320–1331.
Gupta, P., Kulkarni, T., & Toksha, B. (2022). AI-based predictive models for adaptive learning systems. In *Artificial intelligence in higher education* (pp. 113–136). CRC Press.
Hawley, W. D. (2006). *The keys to effective schools: Educational reform as continuous improvement*. Corwin Press.
Jahan, M., & Doggett, M. (2015). *A study on the students' perceptions of the applicability of lean principles at universities*.
Kaswan, K. S., Dhatterwal, J. S., & Ojha, R. P. (2024). AI in personalized learning. In *Advances in technological innovations in higher education* (pp. 103–117). CRC Press.
Kwan, A. (2009). Problem-based learning. In *The Routledge international handbook of higher education* (pp. 91–108). Routledge.
Lavanya, C., & Murthy, J. N. (2022). Assessment and attainment of course outcomes and program outcomes. *Journal of Engineering Education Transformations, 35*(4), 104–111.
Mahon, J. P. (1991). Kaizen: The key to Edu cational improvement. *NASSP Bulletin, 75*(538), 145–148.
Moen, R., & Norman, C. (2006). *Evolution of the PDCA cycle* [online].
Mok, H. N. (2014). Teaching tip: The flipped classroom. *Journal of Information Systems Education, 25*(1), 7.
Park, S., Hironaka, S., Carver, P., & Nordstrum, L. (2013). *Continuous improvement in education*. Advancing Teaching–Improving Learning. White Paper. Carnegie Foundation for the advancement of teaching.
Parsons, C. (Ed.). (2017). *Quality improvement in education: Case studies in schools, colleges and universities*. Routledge.
Pataranutaporn, P., Danry, V., Leong, J., Punpongsanon, P., Novy, D., Maes, P., & Sra, M. (2021). AI-generated characters for supporting personalized learning and Well-being. *Nature Machine Intelligence, 3*(12), 1013–1022.
Pedaste, M., Mäeots, M., Siiman, L. A., De Jong, T., Van Riesen, S. A., Kamp, E. T., Manoli, C. C., Zacharia, Z. C., & Tsourlidaki, E. (2015). Phases of inquiry-based learning: Definitions and the inquiry cycle. *Educational Research Review, 14*, 47–61.
Raghunath, R., Anker, C., & Nortcliffe, A. (2018). Are academics ready for smart learning? *British Journal of Educational Technology, 49*(1), 182–197.
Satterlee, B. (1996). *Continuous improvement and quality: Implications for higher education*.
Sylvie Beatrice, E. (2024). *Personalized learning paths: Leveraging data analytics for tailored education*.
Taylor, R., Fakhimi, M., Ioannou, A., & Spanaki, K. (2024). Personalized learning in education: A machine learning and simulation approach. *Benchmarking: An International Journal*.
Temponi, C. (2005). Continuous improvement framework: Implications for academia. *Quality Assurance in Education, 13*(1), 17–36.
Wiid, P. (2018). Kaizen and education. In *Agile and lean concepts for teaching and learning: Bringing methodologies from industry to the classroom* (pp. 63–92). Springer Singapore.
Zhang, L., Basham, J. D., & Yang, S. (2020). Understanding the implementation of personalized learning: A research synthesis. *Educational Research Review, 31*, 100339.

AI and ML for Continuous Monitoring of Cognitive Learning

R. Kanthavel, R. Adline Freeda, and R. Dhaya

Abstract This chapter explores how Artificial Intelligence (AI) and Machine Learning (ML) can enhance cognitive learning by providing ongoing, data-driven assessments of student progress. Within this domain, it reveals an essential technical gap in conventional education institutions: unlike more clinical environments, no system or classroom structure provides timely and adaptive feedback for personalized learning. In efforts to improve this process, we examine the use of AI technologies, including Natural Language Processing (NLP), Deep Learning (DL), and Predictive Analytics (PA), to observe cognitive performance, identify patterns of learning, and facilitate adaptive responses. This chapter describes the delineation of a structure, encouraging AI-informed alterations to assessment tools, to assess comprehension, engagement, and knowledge retention. To illustrate these approaches, we provide an AI-based application example that can identify cognitive overload or constraints in student learning, enabling timely and personalized support. A case study is presented to exemplify the practical application of these strategies, while fostering equitable learning environments that promote meaningful engagement and informed pedagogical responses. Finally, we consider the ethics of AI and education, keeping a focus on balancing automation and human responsibility to avoid a future of complacency. We discuss the challenges and opportunities for an AI cognitive educational system, in conjunction with the possibilities for future research and recommendations for ethical guidelines, to support this initiative and facilitate its deployment.

Keywords Artificial intelligence (AI) · Machine Learning (ML) · Cognitive Learning · Natural language processing (NLP) · Predictive analytics (PA)

R. Kanthavel · R. Dhaya (✉)
School of ECE, PNG University of Technology, Lae, Papua New Guinea

R. A. Freeda
Department of Information Technology, KCG College of Technology, Chennai, India

D. Vetrithangam et al. (eds.), *Transforming Outcome-Based Education with Machine Learning*, https://doi.org/10.1007/978-981-95-4925-2_5

1 Introduction

Cognitive learning describes the process of acquiring, understanding, and applying knowledge to develop solutions through reasoning. Unlike rote learning, which involves memorizing facts or concepts at a superficial level, cognitive learning entails a deep understanding, critical thinking, and the ability to apply knowledge across unrelated domains of inquiry (Zhai et al., 2021). It represents a higher-level form of learning that arms the learner with the capability to analyze, evaluate, and ultimately synthesize knowledge, therefore creating a deeper learning and cognition experience that is more meaningful and more easily translatable into real-life complex challenges (Chen et al., 2020).

Traditional forms of assessments, such as midterms, finals, and standardized tests, for the purpose of final evaluations, provide insight into a student at one moment, usually within a specific time frame. While these forms of traditional assessments are very prescriptive in being standard and provide clarity as to what is seen as important content of study in a conventional sense, they do not accurately reflect the ever-changing process of cognitive development of students (Tempelaar et al., 2015). The traditional forms of assessments often miss important pieces of cognitive process by not capturing some of the behavioral processes of organizing, engaging, and adhering to knowledge over a period of time. As the world evolves and shifts toward a digitally driven ecosystem, education should evolve and shift from prescriptive, rigid, ethnocentric, and one-size-fits-all approaches to a personalized, adaptive, learner-centered continuum of the learning process that meets learners where they are at within their learning styles and needs (Bhatt et al., 2023).

The increasing demand for personalized, data-informed learning solutions has resulted in continuous cognitive monitoring, which allows educators to assess a student's cognitive state, levels of engagement, and comprehension in real time. Continuous monitoring allows educators, students, and researchers to monitor progress, track learning gaps, and provide targeted intervention strategies to support learning. Continuous monitoring more broadly supports learners who need individualized attention to optimally meet potential learning outcomes, particularly in our fast-paced world of knowledge, where assessments may not necessarily predict outcomes or learning needs. AI and ML are shaping cognitive learning through intelligent monitoring and automated learning solutions that analyze and draw insights from educational data to assess learning patterns and provide feedback to learners (Javed et al., 2023). AI-infused systems may provide an assessment of moments of engagement and disengagement for a student, signal cognitive fatigue, or predict a potential learning barrier before the student becomes stalled in the learning process. These AI-powered solutions may include NLP, DL, and PA within a personalized learning experience that adapts and responds based on the learner's individualized needs. AI-based cognitive learning systems can also support inclusive learning practices by providing differentiated learning experiences to meet diverse learning needs, provide interventions during learning moments and also be a scalable solution in educational institutions and e-learning platforms.

This chapter looks at the importance of continuous cognitive monitoring in education today, how AI and ML can change traditional learning approaches, and the promise of intelligent algorithms in creating more personalized and effective learning experiences. Finally, we discuss ethical questions associated with AI in education, addressing the need to balance automation with the human ability to monitor and assess learning to ensure responsible and beneficial implementation of AI-enhanced learning systems.

2 The Challenges of Traditional Monitoring Approaches

Cognitive learners are monitored traditionally by using assessment of standard tests, quizzes, and teacher evaluations. Although these methods of assessing learning have been widely used in schools, they are accompanied by a number of limitations that thus bar them from offering real-time, individualized information about a learner's cognitive learning (Wei et al., 2021). Figure 1 shows the Limitations of the standardized framework.

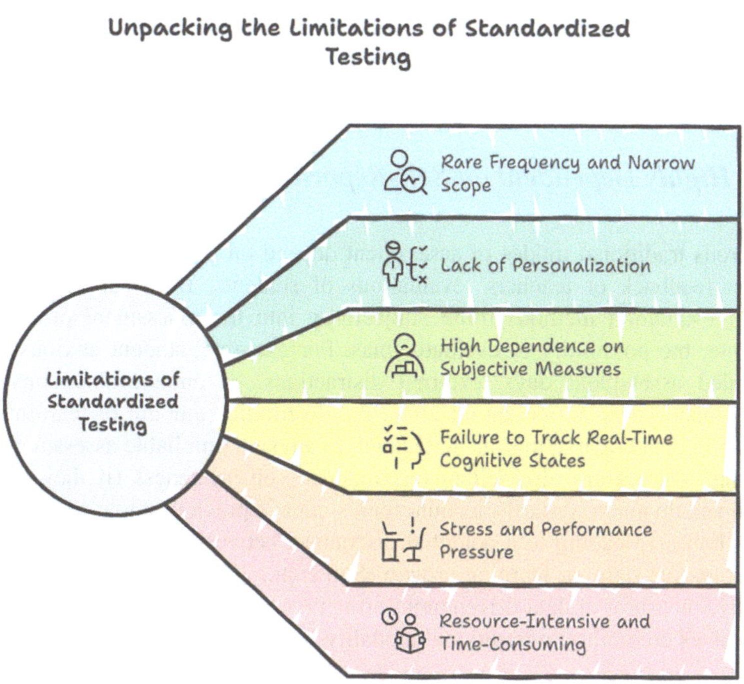

Fig. 1 Limitations of the standardized framework

2.1 Rare Frequency and Narrow Scope

Standardized assessments are provided periodically, giving a glimpse of a student's growth. These periodic assessments do not account for the daily variations in cognitive performance, learning behaviors, and comprehension (Halkiopoulos & Gkintoni, 2024). Because of this, both teachers and students are missing out on weaving in timely remedial instruction, which may leave gaps in understanding for long periods of time depending on the assessment system's schedule. Real-time assessment of learning diminishes the purpose of regular evaluations in that they do not measure student learning continuously and validate cognitive growth (Ahmad et al., 2024).

2.2 Lack of Personalization

Every student engages in their respective cognitive processes, at their own pace, and in their own manner. Standardized assessments and assessments of learning do not account for these variances; thus, these processes yield blanket measures that fail to measure the strengths and weaknesses of all students. These assessments, and most often school processes and programs, provide students with universal choices and options that may or may not fit their unique learning needs. Furthermore, when adaptive learning models are absent, there is more student disengagement and little chance for students to learn successfully while in school.

2.3 Highly Dependent on Self-Reporting and Subjective Rating

Numerous traditional modes of assessment depend on some type of self-reported student feedback or teachers' evaluations of students. Both self-reporting and teacher evaluation methods bring subjectivity into the assessment process and, therefore, the possibility of evaluator bias. For example, student anxiousness on scheduled assessment days, external distractions, or unbelievable consistency among teacher evaluators can all affect the determined amount of learning in an assessment, leading to a miscalculated and, as a result, unreliable assessment. Self-reporting sources are also reliant on students' self-awareness (if they even are aware) and do not necessarily account for a student's previous knowledge of a topic and/or honesty, leading to a potential disconnect between what they reported and their cognitive state, or learning growth (Sun et al., 2024). Again, the inclusion of objective measures or data-driven monitoring processes across instructional periods undermines credibility in traditional modality-based assessments.

2.4 Failure to Track Real-Time Cognitive States

Conventional assessments are only concerned with static knowledge retention instead of cognitive functions taking place in real time, such as attention, retention capacity, and problem-solving skills. They do not offer insight into a learner's cognitive load, engagement, or difficulties as they occur, meaning educators cannot intervene at the appropriate time to minimize potential learning disruptions and provide timely support for learners. A lack of real-time data means the educator cannot take any proactive educational actions; therefore, they cannot optimize their educational interventions.

2.5 Stress and Performance Pressure

Frequent high-stress assessments can create anxiety and stress in students, which can impact student outcomes. The expectation of achieving a high score can lead to cramming, shallow learning, and reduced retention of knowledge for the long term. Performance anxiety can compromise the integrity of the test, leading to a false portrayal of the student's cognitive capacity. Instead of allowing for deeper learning of concepts, traditional testing may contribute to a stressful learning environment that doesn't support student cognitive development (Wei et al., 2024).

2.6 Resource-Intensive and Time-Consuming

Standard assessments can take considerable time and effort to manually grade, create feedback, and deal with all administrative pieces, which is a luxurious diversion of time and attention. Teachers place significant emphasis on grading individual performance, delaying educator feedback and intervention. This is particularly complex for very large classrooms or online educators, where even more personalized attention is difficult to provide (Li et al., 2024). Relying on human evaluators for grading is not scalable, nor timely, nor consistent, and severely limits the ability to provide timely, informative, data-led instructional improvement.

2.7 Inflexible Adaptation to Novel Learning Environments

As digital and remote learning become more prevalent, standard monitoring techniques begin to fall short of the overall trend in education technology. In online and self-paced learning environments, there is more of a need for real-time monitoring and automated assessments for measuring progress. Traditional evaluation systems cannot incorporate the needed flexibility to merge with educational technology, contributing to a lack of information about the dynamic and changing nature of digital education. The larger movement toward AI platforms represents a call for new monitoring techniques that may be automated and adaptive (Zhang & Heffernan, 2024).

AI and ML-based cognitive monitoring systems correct many of the limitations in traditional assessments by offering a consistent, adaptive, and data-oriented approach to assess the learning process. These systems lean on real-time analytics, predictive modeling, and personalized feedback to increase the learning experience (Shi, 2019). AI-based systems can continuously monitor student performance, detect cognitive overload, and take timely actions. Through the automation of assessments and consideration of needs for learning strategy adjustment for specific individuals, AI and ML will facilitate a more effective and inclusive education system.

3 How AI and ML Can Transform Cognitive Learning Monitoring

The use of AI and ML in education is changing the way that cognitive learning is assessed and optimized. Traditional methods of learning assessment, which include word problems and standardized tests, do not capture real-time variation in the learner's cognitive engagement, memory retention, or problem-solving ability. With AI and ML, there are data-driven, adaptive systems that can continuously monitor patterns of learning, identify cognitive fatigue, and give feedback to improve the learning process. AI cognitive monitoring systems utilize complex technology, such as NLP, PA, and real-time monitoring, to provide insight into the learner's cognitive state (Lee et al., 2019).

Traditional approaches only provide information about what knowledge has been learned at specific intervals, whereas AI-enabled systems enable continuous monitoring and to intervene in a timely manner and create authentic learning experiences. Intelligent systems can detect disengagement, monitor understanding, and customize or elevate the difficulty of the learning content based on a learner's needs. AI is also encouraging educational equity by providing different approaches that consider a learner's style and levels of ability, facilitating equitable access for the learner. In combination, cognitive monitoring, automated assessments, tracking behavioral patterning, and automated recommendations allow a dynamic and responsive learning experience that will create the learning space of the future. Continuous, real-time monitoring of cognitive learning with integration with AI and ML can be used to revolutionize education as follows: the ability to dynamically analyze learning behaviors, make predictions, and provide adaptive feedback tailored to individual needs, unlike traditional methods of monitoring (Howard, 2019).

3.1 *Real-Time Cognitive Monitoring*

AI-based tools assess learners in their interaction with digital content by tracking response times and analyzing problem-solving behaviors. Technologies like eye-tracking, speech recognition software, and sentiment analysis capture cognitive load in-the-moment, engagement, and attention levels. Data collected through these processes provides educators with awareness so they may intervene (when appropriate) or adjust teaching strategies to improve learner effectiveness or outcomes. AI is able to monitor cognitive load closely, thereby improving learning experiences with ongoing, real-time educational or personalized learning experiences.

3.2 *Personalized Learning Experience*

ML algorithms analyze large amounts of student data to uncover individual learning styles, strengths, and weaknesses. Using artificial intelligence, platforms develop individualized study plans, change content difficulty, and offer individualized resources to improve the learning experience. This personalized approach guarantees that students have relevant study materials that promote engagement and foster deeper learning and retention. When AI personalizes the learning experience to suit the needs of each student, it results in a more effective and student-centered approach to learning.

3.2.1 NLP Deeper Cognition

NLP methodologies allow AI systems to read, understand, and observe language in written or spoken form, assessing reasoning, critical thinking, and comprehension abilities. Specifically, AI-based virtual tutors and chatbots represent new kinds of educational interlocutors, engaging learners in discourse that includes follow-up questions to their responses that reinforce/consolidate conceptual understanding. The use of NLP to provide these human-like tutoring experiences creates cognitive learning because the educational experience is dynamic, responsive, and provides educational interaction cognitively represented as dialogue (Liu et al., 2024).

3.2.2 AI-Based Learning Analytics

AI-powered dashboards provide educators and institutions with insight into the trends of student performance, how students engage, and what they retain. These analytics aid in the design of evidence-based learning strategies and curriculum improvements.

3.2.3 Intelligent Feedback and Suggestions

An ML model learns from past learning behaviors and recommends personalized resources in the form of videos, articles, and exercises that a student finds relevant to their learning pace and interests. This boosts engagement and retention.

3.2.4 Monitoring of Cognitive and Emotional State

The AI-based emotion recognition and sentiment analysis can capture the frustration of a learner, motivation, or confusion through facial expressions, voice tone, and keystroke patterns. So, emotional intelligence gets integrated to develop a more responsive learning environment.

AI and ML technologies redefine the process of cognitive learning monitoring as more personalized, proactive, and efficient. They empower educators and learners with real-time analytics, adaptive learning models, and predictive insights that enable them to offer more effective and engaging educational experiences. However, data privacy, bias mitigation, and human-AI collaboration need to be addressed to maximize the benefits of AI-driven learning systems.

4 Key Applications of AI and ML in Cognitive Learning Monitoring

AI and ML technologies have transfigured cognitive learning monitoring by allowing constant, personalized, and data-driven assessment. These applications improve engagement among students and outcomes while making the job of educators more valuable. Applications of AI and ML in Cognitive Learning Monitoring are shown in Fig. 2. Some of the key applications of AI and ML in cognitive learning monitoring are as follows:

4.1 Intelligent Tutoring System (ITS)

AI-based tutoring systems adapt their instruction to every student's learning pattern, pace, and performance. It personalizes the information and exercises, explains, and provides feedback based on a student's performance to better understand (Elallid et al., 2022). Examples of these are AI-driven platforms such as Carnegie Learning's MATHia and IBM's Watson Tutor.

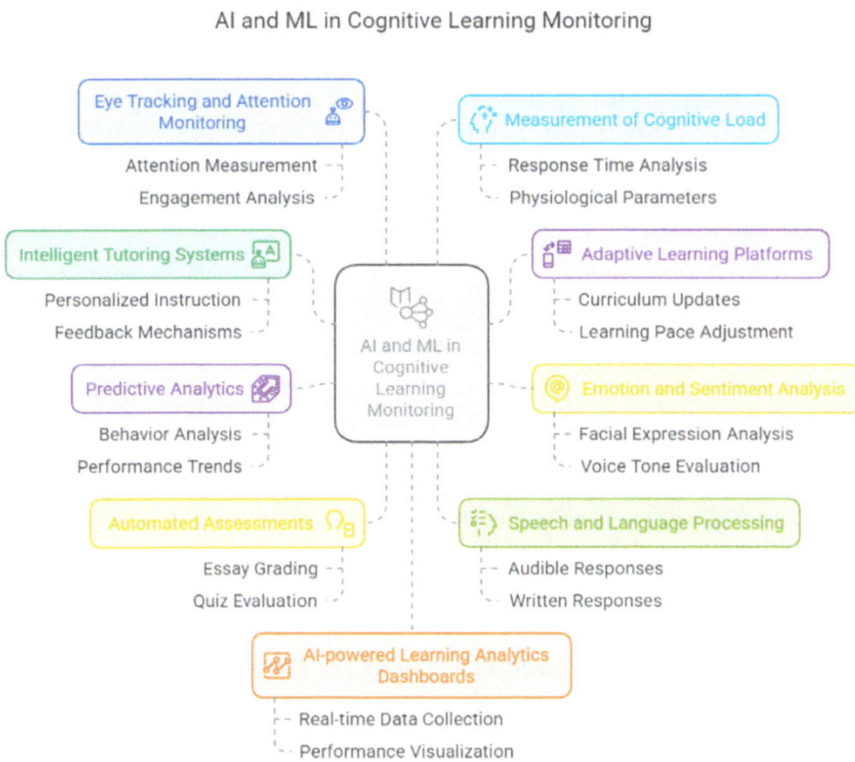

Fig. 2 Applications of AI and ML in cognitive learning monitoring

4.2 Adaptive Learning Platforms

ML algorithms learn from the performance of learners and update the curriculum accordingly to adjust to the learning pace of students. Adaptive learning in Coursera, Duolingo, and Knewton is an application that displays content according to a learner's requirements for optimum learning.

4.3 Emotion and Sentiment Analysis for Tracking Engagement

AI can analyze facial expressions, voice tones, and text responses. Such records will make it easy for an instructor to know whether a student is frustrated, bored, or confused. The instructor will, therefore, adjust the teaching strategies accordingly. Some of the tools that utilize AI in measuring the emotional state of learners include Affective and Cognimates.

4.4 PA for Early Intervention

ML-based models predict the students likely to fall back by analyzing the learning behavior of such students, performance trends, and engagement level. The early warning signs can be picked up to intervene before there is a possible setback in learning (Chatterjee, 2020). PA is most applied in Brightspace Insights and Blackboard Predict.

4.5 Automated Assessments and Feedback

These automatically grade essays, quizzes, and assignments with the help of NLP and ML. Such assessment tools give an immediate response to mistakes to students so that the students can know their mistakes, so that they won't wait to learn from these mistakes when graded by humans. Some popular assessment solutions using AI are: Turnitin, GradeScope, and WriteLab.

4.6 Speech and Language Processing for Cognitive Assessment

NLP makes it possible for AI to judge the audible and written responses given by students in order to ascertain their understanding, logical thinking, and critical ability.

4.7 Eye Tracking and Attention Monitoring

AI-based eye tracking technology determines how students' attentions are applied in learning, what to measure, and what they look for, as regards the extent of engagement with a particular cognition load. More so, in optimizing instruction in digital learning settings and improving digital learning user interfaces (Sangeetha et al., 2021).

4.8 Measurement of Cognitive Load

ML algorithms check out the cognitive overload in relation to response time, keystroke pattern, and also physiological parameters of students' bodies, including the heart rate variability. The outputs are employed in order to calibrate learning by eliminating every overload that, beyond a particular span, triggers amnesia.

Brainwave tracking sensors from wearables like NeuroSky and Emotive back these AI-driven solutions for measuring the cognitive overload.

4.9 AI-Powered Learning Analytics Dashboards

AI-based learning analytics platforms collect information about student performance, engagement, and progress in real-time, and it is visualized. These data help educators make informed decisions while improving curriculum designs. Some examples of this are Power BI for education and AI-driven analytics through Google Classroom.

4.10 Gamification and AI-Driven Learning Experiences

AI is combined with gamification techniques to engage the learners and sustain motivation. AI-created challenges and adaptive rewards on personalized learning tracks encourage learners to achieve active participation in learning (Xu, 2024). The following are some of the platforms that apply AI-driven gamification strategies: Kahoot! and DreamBox. From the monitoring of cognitive learning with AI and ML, applications go beyond this old assessment. AI represents good representations of smart tutoring, adaptive learning, emotion recognition, PA, and real-time feedback that would make learning more personal and effective. As defined in the development process so far, they will show the definition of education in the future—making it easier to access, interact, and be informed by data.

5 Personalized Feedback and Adaptive Learning Experiences

In conventional education systems, feedback and learning effects are generally standardized for all students and rarely take into consideration the unique needs of the students. Standardized tests (or termly assessments) give learners feedback that is delayed and do not enable learners to rectify mistakes or reinforce concepts in real time. Furthermore, conventional pedagogies often do not account for students' variable cognitive abilities, learning trajectories, or interests, which often leads to disengagement and ineffective learning. AI and ML are, evidently, going to change this by facilitating personal feedback and personalized learning experiences. AI educational tools assess students' learning activities to understand their learning patterns and provide immediate and personalized feedback to help the learners' understanding. These systems provide transformation and continuously change the difficulty of the content or provide personalized recommendations, and potentially optimize instructional practice for learnings' unique cognitive profile forms. Using a range of technologies, including NLP, sentiment analysis, adaptive tutoring systems, and real-time analytics, AI learning platforms develop more responsive and engaging learning experiences. AI instances in education range from grammar recommending applications, like Grammarly, to adaptive learning systems like Knewton and DreamBox, which directed learners through optimized learning pathways for retention and cognitive growth.

5.1 AI and ML-Based Customized Feedback Mechanisms

Standardized assessment systems usually have the characteristic of delayed evaluations, thus preventing them from actually supporting the learning improvement in real time. This is corrected with AI-based feedback mechanisms by delivering immediate insights driven by data. Learning systems powered by AI analyze student responses and offer immediate feedback so that students may rectify errors or reinforce learning at the point in time (Sathyaraj et al., 2022). ITS, such as Carnegie Learning's MATHia, explain each step based on the student's response. NLP tools grade essays according to grammar, coherence, argument strength, and originality based on written responses. Grammarly, WriteLab, and Turnitin use AI to give writing quality feedback and plagiarism detection.AI-based sentiment analysis picks on frustration, confusion, or even boredom via facial expressions, tone of voice, and interaction patterns. The adaptive systems adapt what is to be delivered with respect to the state of emotions; encourage or simplify the issues as they see fit. AI detects learning behaviors and generates adaptive suggestions for studying, resources, and activities which cater to some specific strengths or weaknesses.

5.2 AI Adaptive Learning

Adaptive learning employs AI and ML, making the instructional content and presentation dependent on learners' cognitive skills, pace of learning, and interest. ML algorithms learn from student performance to adapt the level of difficulty, the speed, and the type of lesson. Adaptive learning systems, such as Knewton and DreamBox, adapt instruction for each learner toward effective learning progress. The AI systems decide the depth of comprehension and also phrase difficult thinking into easier modes or even multimedia models.AR and VR engage students in interactive adaptive learning environments.AI uses game-based strategies, among them making available tailor-made challenges together with rewards. These make the learner more interested and also trigger cognitive recall. Duolingo and other similar apps adjust the level of their content difficulty according to the learner's performance. As such, the level of motivation from learning will continue improving. AI-driven chatbots and voice assistants deliver instant, interactive, and on-demand tutoring services in the form of answering questions or helping the learner to a solution of the problem. Few instances include IBM Watson Tutor and Google's Read Along which provide tailor-made reading aid.

5.3 Benefits of AI-Based Personalized Learning

AI and ML have numerous opportunities for education. Using themed content and gamification, they help keep students engaged and make learning more fun. Dynamic learning makes sure that learners need to reinforce the conceptual learnings based on their learning requirements leading to better retention. AI time management is about giving the student the right amount of time while omitting those that the student has mastered to save time and move quickly. By adapting to the student's pace, it helps lessen the cognitive overload and prevents the student from feeling overwhelmed. AI also helps make learning more accessible as it caters to multiple learning styles and learning strengths. Education becomes more effective, interactive and suited to the needs of each student through AI and ML, which are used to give personalized feedback and to adjust lessons. The result is a sustainable future of learning that will take place in classrooms and virtually that is responsive and supportive to help all students move forward successfully (Rohit et al., 2024).

6 Implementation of Cognitive Learning

Cognitive learning is a process which is centred on knowing and applying information, and not just memorizing facts. It calls for active engagement, critical thinking, problem-solving, and the power to relate new knowledge with existing experience. Using cognitive learning needs careful strategies which involve learners' engagement, prompting reflection, and facilitating application in real life. Foremost, there must be an engaging learning environment set. This comes in the way of lessons tailored to bring forth curiosity and intellectualism (Sakal et al., 2024). The instructor can include students in lively techniques like debate, discussion group, and questioning time to create a critically thinking mind from information. Placing learners into life-like settings, case study or problem-solving class may facilitate that students are doing what they do know apply them. For instance, when there is science class, the students can be utilized to produce real-life problem-solving solutions, e.g., creating an energy system that can be sustained. This method allows them to become more critical and comprehend the topics better.

Encouraging active learning is also an effective technique. Cognitive learning is optimally done when learners are actively doing it. Brainstorming activities, mind mapping, and concept mapping enable students to structure their thoughts and relate new information to prior knowledge. For example, once a new idea has been learned, students can be required to produce a mind map illustrating how the idea fits into what they already know. These activities consolidate learning and facilitate recall later. Furthermore, the use of questioning in class makes students think and build critical thinking. Visual and experiential tools are also another important application of cognitive learning. Visual tools like charts, diagrams, videos, and infographics make complicated concepts simpler and easier to understand. Demonstrations and hands-on exercises make students see and do things for themselves, making observing and practicing to understand more potent. For instance, in mathematics class, teachers can employ visual images to symbolize abstractions like fractions or geometric figures. Likewise, field observation and experiments in science enable learners to experience theoretical concepts, leading to higher cognitive processing and remembering (Al-Hindawi et al., 2025).

6.1 Selection of ML Algorithm for Cognitive Learning

Choosing a ML algorithm to use in cognitive learning. To enable cognitive learning, we may use reinforcement learning (RL) with deep neural networks or supervised learning with PA. As an example, some of the most common cognitive learning represented algorithms in ML. RL makes it possible to design adaptive learning pathways that tailor lessons to match each learner's pace, providing them instructions that are tailored for their individual experience. Key Uses of DL Nikola Tesla, naturally enough, could use to NLP, DL is used in most NLP technologies, allowing systems to continue giving personalized feedback and responsive learning solutions. Decision Trees and Random Forests will help us identify the students who need more help so early we can intervene. Support Vector Machines (SVM) plays an important role in monitoring the emotional state and engagement level of students using sentiment analysis. Finally, Bayesian Networks are applied to e-learning environments, specifically intelligent tutoring systems, to provide personalized learning recommendations based on student performance (Mohamed, 2023).

6.2 AI Driven Cognitive System Workflow

Cognitive Learning System powered by AI that is comprised of the below in workflow Fig. 3: Here is the architecture diagram of an AI-driven Cognitive Learning System. It illustrates how data moves from student interactions through preprocessing, model training, personalized feedback, and evaluation to enhance learning experiences.

This layer is designed to collect various forms of data related to student engagement (quiz scores, interaction logs, behavioral patterns, and facial expression data to serve emotional signatures). This raw data is finally forwarded to the Preprocessing

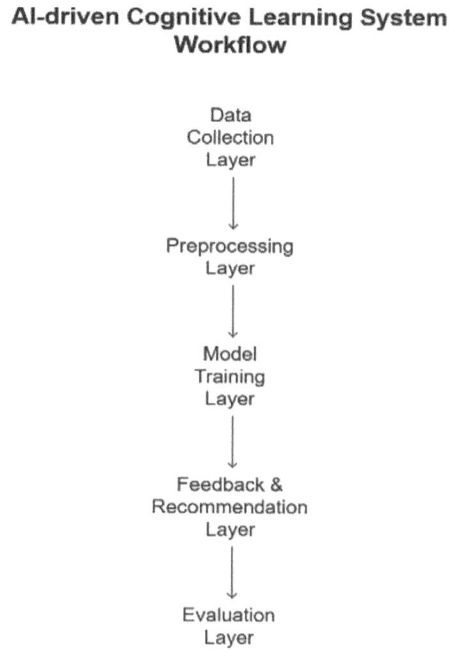

Fig. 3 AI driven cognitive learning system workflow

Layer, where NLP and DL techniques are used to clean, normalize, and formulate meaningful features that indicate students' learning behaviors and cognitive patterns. The purified data is used by the Model Training Layer, which employs RL and PA methods to create adaptive learning models that can understand personalized learning paths. We also fine-tune the learner engagement and cognitive patterns based on the data mentioned earlier from Model Output & provide a Feedback & Recommendation Layer to give real-time personalized learning outputs. The Evaluation Layer constantly evaluates student progress and performance, adjusting their learning trajectories as it detects the underlying structure of the material being assimilated to optimize the retention of knowledge over the long term.

6.3 *Implementing the ML Model*

To implement a Neural Network-based Adaptive Learning Model, we integrate student engagement data (quiz scores, engagement levels, and learning speed) with RL and PA (Random Forest, SVM) for cognitive monitoring. Additionally, NLP enhances personalized feedback by analyzing written responses and engagement patterns. The model is trained on synthetic student performance data to predict learning outcomes dynamically. The neural network adapts content difficulty based on real-time student interactions, while RL optimizes personalized interventions to improve learning retention. Model evaluation includes accuracy, precision, recall,

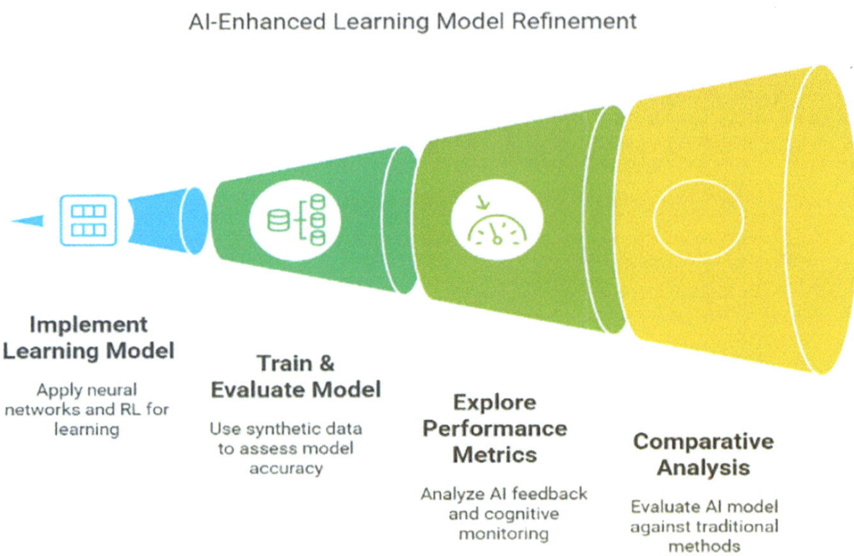

Fig. 4 Implementing the ML model

and F1-score to compare AI-driven cognitive monitoring against traditional assessment methods. Results show higher accuracy, real-time adaptability, and personalized feedback, addressing cognitive overload and learning gaps proactively (Tawiah, 2022). Comparative analysis demonstrates that AI-powered models outperform traditional methods by offering real-time insights, adaptive learning paths, and early intervention capabilities, ultimately leading to improved student engagement and retention. Figure 4 explains the implementation of the ML model.

6.4 Results of AI-Based Cognitive Learning Monitoring

The following parameters helped to find the Results of AI-based cognitive Learning Monitoring.

- **Model Accuracy**: The Random Forest model achieved 60.5% accuracy in predicting adaptive learning success.
- **Accuracy**: The proportion of correct predictions made by the model compared to the total predictions. It is a general measure of how well the model performs.
- Precision (Positive Predictive Value): The percentage of true positive predictions out of all predicted positives. It indicates how accurate the model is when it predicts that a student belongs to a certain class. For students needing improvement (class 0), Precision = 46%. This means that when the model predicts a student needs improvement, 46% of the time, it is correct.

Table 1 Comparative analysis

Aspect	Traditional Assessments	AI Models (Random Forest)	DL Models (LSTMs, Transformers)
Real-time Learning trends	Fail to capture real-time trends.	Can dynamically analyze real-time learning patterns.	Can process sequential data and adapt to trends more effectively.
Feedback	Limited feedback after assessment.	Continuous feedback based on real-time performance.	Provides more accurate and adaptive feedback.
Data processing	Based on fixed assessments (e.g., exams, quizzes).	Analyzes continuous data for more flexible learning insights.	Better at handling large volumes of data with complex patterns.
Accuracy	Often static, limited to periodic evaluations.	Can achieve higher accuracy in identifying success or failure.	Potential for greater accuracy due to deeper analysis of patterns.
Adaptability	Less adaptive to individual learning styles and needs.	Adaptable to different learning patterns in real-time.	Highly adaptive, continuously improving predictions over time.
Future potential	Limited by traditional methods of evaluation.	Moderate improvement with enhanced data analysis techniques.	Significant potential for improvement in learning prediction accuracy.

- **Recall (Sensitivity or True Positive Rate)**: The percentage of true positive predictions out of all actual positives. It measures how well the model identifies all students who actually belong to a certain class. For students needing improvement (class 0), Recall = 35%. This means that the model is only able to correctly identify 35% of the students who need improvement.
- **For successful learners (class 1)**: Precision = 66% and Recall = 75% The model performs better at identifying students who succeed but struggles with those needing improvement.

Here is a Table 1 explains the comparative analysis. While traditional assessments offer limited, static feedback, AI and DL models offer real-time, dynamic insights with growing potential for accuracy and adaptability in understanding students' learning processes.

Finally, for cognitive learning to be implemented, a multidimensional strategy that involves active teaching styles, active engagement, reflective practices, the use of technology, and ongoing feedback is essential. Through these mechanisms, teachers are able to promote greater understanding, critical thinking, and effective application of knowledge (Alowais et al., 2023). This strategy prepares learners for real-life problems by giving them skills that transcend the classroom.

7 Case Studies and Effective Implementation Strategies

AI and ML have significantly transformed the educational landscape, especially in cognitive learning monitoring. By offering personalized feedback, adaptive learning, and dynamic assessment, these technologies allow for more tailored and efficient learning experiences. Below are several case studies showcasing the effective implementation of AI and ML in education, followed by key strategies for successful adoption. Each case study operationalizes OBE principles by embedding ML methodologies into assessment, personalization, and intervention processes. For instance, Carnegie Learning's MATHia uses supervised ML assessment models to adapt problem difficulty in real time, ensuring mastery-oriented progression. Duolingo integrates NLP with reinforcement learning algorithms to analyze pronunciation and dynamically balance lesson difficulty, thereby supporting OBE goals of skill mastery. IBM Watson employs ML-driven dialogue and predictive models to deliver personalized tutoring, real-time feedback, and adaptive study plans. Georgia State University applies predictive analytics algorithms to student records for early identification of at-risk learners, enabling proactive interventions aligned with outcome achievement. Similarly, Knewton utilizes adaptive ML recommendation systems to personalize learning trajectories and provide data-driven dashboards that support continuous OBE assessment.

7.1 Case Study 1: Carnegie Learning—AI-Powered Math Tutoring

Carnegie Learning's MATHia software uses AI to provide personalized math education for middle and high school students. It dynamically adjusts based on the students' responses and guides them through the learning process. If a student struggles with solving algebraic equations, MATHia will offer additional practice in this area, with hints tailored to the student's preferred learning style (visual, auditory, etc.), which significantly improves mastery of the topic is a real time example. The two impact values are 30% improvement in math proficiency in schools using MATHia, and Higher student engagement and increased confidence in problem-solving.

Implementation strategies are another way that MATHia and other similar platforms help personalize and tailor learning I One example is called ML assessment, in which software evaluates student answers in real time and modifies problem difficulty based on performance. It also provides a step-by-step strategy for building instruction around students' unique ways of learning, which better promotes understanding. Also, MATHia produces live reports on how students are performing that teachers can use to monitor which students are doing well and who might need additional help. Then teachers are able to guarantee that support at the exact time and place it's required.

7.2 Case Study 2. Duolingo—Adaptive Language Learning

Duolingo uses AI to personalize language learning for users worldwide by adapting lessons based on proficiency and engagement. If a user consistently makes mistakes with verb conjugations, Duolingo will provide additional exercises to reinforce this concept. Gamification features like earning rewards for correct answers keep the learner motivated and engaged, which is a real-time example. The two impact values are 500 million users worldwide and Language skills equivalent to one semester of university study after just 34 h of use. Duolingo is an example of an app that uses several intelligent features in its implementation strategies to ensure that learning takes place. This is NLP, and it allows the app to evaluate the user's pronunciation and grammar while providing corrective feedback. Dynamic difficulty adjustment is another one; the application modifies the difficulty level depending on the user's accuracy and the time spent answering each question. This makes sure the learning process stays balanced—not too easy and not too hard. Gamification is also an effective strategy to make learning more enjoyable and challenging by incorporating rewards, challenges, and leaderboards, which help users stay consistent and motivated to learn.

7.3 Case Study 3. IBM Watson—AI Virtual Tutors

IBM Watson is integrated into several educational platforms to offer AI-driven tutoring and cognitive learning tracking. If a student asks Watson about a complex math concept, Watson uses its database and NLP capabilities to provide a detailed answer, breaking it down step by step. It can then recommend additional practice problems or videos for further understanding, which is a real-time example. The two impact values are increased student understanding and lower dropout rates, and Enhanced efficiency of educators by streamlining grading and assessment tasks. Tracks things in learning, etc. (NLP, Machine learning, and more). For instance, IBM's Watson utilizes NLP to communicate with students, respond to their questions, and provide them with personalized assistance. It also examines the students' past patterns of learning to tailor study plans that suit their requirements, allowing them to spend time on concepts where they require additional practice. Real-time feedback is yet another of the most crucial features; this answer gives students instantaneous feedback on their submissions and recommends additional learning resources for practice and improvement. These tools can make education highly personalized, effective, and efficient.

7.4 Case Study 4. Georgia State University—AI for PA

Georgia State University implemented an AI-based early warning system to detect students at risk and intervene before problems escalate. If a student is showing a decline in grades or has stopped attending classes, the AI system can automatically trigger an alert to counselors, who then reach out to provide support in a real-time example. The two impact values are reduced dropout rates by 22% and increased graduation rates by 6%. Strategies for implementation include the deployment of PA and automated chatbots to assist students. PA analyzes data on student engagement, grades , and attendance to identify the students who are likely to be struggling and need additional support. This enables teachers or school staff to intervene earlier and offer support. Other useful assistants, such as an automated chatbot like "Pounce," help students by responding to their questions about an academic or administrative task.

7.5 Case Study 5. Knewton—AI-Driven Adaptive Learning in Higher Education

Knewton provides adaptive learning technology that customizes coursework for students in universities, particularly in math and science. A student struggling with calculus might be provided with additional resources such as practice problems, video tutorials, and step-by-step guides, all tailored to their specific learning needs and pace is the real time example. Two impact factors are a 55% improvement in student performance in math and science courses and increased completion rates for online learning modules. Here are some of the top education-focused AI implementation strategies with helpful tools. ML algorithms help tailor each student's learning journey according to their strengths and weaknesses, placing them at an appropriate level of support. Another is personalized content, where platforms like Knewton recommend educational resources based on the student's current level of progress, enabling them to learn at their own pace. AI dashboards are also super helpful for teachers since they produce clearer reports about how students are performing. This enables educators to adjust or fine-tune lesson plans to ensure they meet each student's needs.

Here, Table 2 summarizes the Case Studies for AI and ML in cognitive learning monitoring.

Table 2 Case Studies for AI and ML in cognitive learning monitoring

Case Study/ Strategy	Description	Implementation Strategy	Impact
Carnegie Learning—AI-powered math tutoring	AI-powered math tutoring system for middle and high school students, offering personalized problem-solving guidance.	– ML adjusts difficulty based on student responses. – provides step-by-step guidance tailored to each student's learning style. – real-time performance reports for educators.	– 30% improvement in math proficiency. – higher engagement and confidence in problem-solving.
Duolingo—Adaptive language Learning	An AI-driven language learning app that personalizes lessons based on user proficiency and engagement.	– uses NLP for pronunciation and grammar analysis. – dynamically adjusts lesson difficulty based on accuracy and time spent. – gamification features like challenges, rewards, and leaderboards for motivation.	– over 500 million users worldwide. – AI-driven personalization increases retention rates. – equivalent to one semester of university language skills after 34 hours.
IBM Watson—AI virtual tutors	An AI-powered tutoring system is used on various educational platforms to provide real-time feedback and personalized study plans.	– uses NLP and ML to interact with students. – provides personalized study plans based on past learning patterns. – suggests supplementary resources and gives real-time feedback.	– increased student understanding. – lower dropout rates in pilot programs. – streamlined grading and assessment for educators.
Georgia State University—AI for PA	An AI-based early warning system to predict students at risk and intervene proactively to support them.	– PA to track student engagement, grades, and attendance. – AI identifies students needing intervention and triggers counseling. – automated chatbot guides students through administrative and learning issues.	– 22% reduction in dropout rates. – 6% increase in graduation rates.
Knewton—AI-driven adaptive Learning	Adaptive learning technology for universities that tailors coursework to individual student needs.	– ML adjusts learning paths based on student strengths and weaknesses. – personalized content recommendations improve engagement. – AI dashboards help educators track progress and optimize lesson plans.	– 55% improvement in performance in math and science courses. – increased completion rates for online learning modules.

8 Effective AI Implementation Strategies in Cognitive Learning Monitoring

AI and ML are having a significant impact on education by improving the monitoring of cognitive learning. The strategies below will focus on utilizing AI to enhance learning outcomes, respond to student needs, and give educators real-time insights. Each strategy is detailed, along with real-time examples of implementation.

8.1 Real-Time Data Collection and Analysis

Real-time data collection and analysis are crucial to understanding a student's engagement, learning behavior, and cognitive load. An AI-based Learning Management System (LMS) tracks interactions, responses, and behaviors to inform how students are performing in the moment. AI-based LMS enables active responses to student engagement, when warranted, to adjust lesson difficulty accordingly or to provoke targeted support. There are a few very useful use cases for AI in education. One way to do that is through artificial intelligence-assisted Learning Management Systems (LMS), which record how a student uses the platform, how much time they spend on tasks, how often they get a given answer correct or incorrect, and how much they interact with the material. This real-time data helps understand how well a student is learning and what they need help getting better at. The other approach is biometric and sentiment analysis. Technologies such as facial recognition and eye-tracking can verify a student's feelings and attention span. This allows a system to adapt immediately to improve learning by seeing whether a student is engaged, frustrated, confused, or distracted. An example is with a platform like Knewton, where a teacher knows how much time a student spends on each question, how many tries a question takes, and even if they are confused by webcam or wearable devices. If a student appears to be frustrated or loses attention, the system can reduce the level of challenge or switch the learning material. AI used for emotion detection can even detect facial expressions, such as when someone is frowning or looking away, and can recommend breaks or less difficult tasks to enable students to stay focused and learn better.

8.2 Models for Personalized Adaptive Learning

Personalized adaptive learning refers to AI systems that customize the educational materials for the unique needs and pace of each student. ML algorithms dynamically modify the difficulty, content, and path of learning to what the student is doing, allowing each and every learner to have a personalized experience. AI can be used in a variety of ways to enhance the experience of learning. One approach involves using machine-learning algorithms that analyze how students answer questions and their performance. The system can use this to adjust the difficulty of tasks, the order in which lessons are presented, or the type of materials provided. Consider a student who excels in geometry but struggles with algebra: the system will assign him more algebra practice, thus allowing him to improve. AI-driven tutoring is another method. Manufactured intellect chatbots run on NLP can provide instantaneous, customized responses for assignments and inquiries. Chatbots guide the students gradually, like a real tutor. In practice, an example is DreamBox Learning, where when a student answers incorrectly, the AI adjusts the question difficulty and offers hints or illustrations tailored according to the listener's learning style so that the student understands and doesn't get frustrated.—the Socratic app from Google, for example. Students go in with tough questions for what they're learning in math or science, and the AI splits the question up into smaller components, giving hints until the student arrives at the answer." These instruments enable personalized, guided, and clearer learning.

8.3 PA for Early Intervention

PA is a relatively new method of utilizing data to identify students who may be at risk of disengaging or dropping out. In this process, a combination of attendance, grades, participation, and behavioral data is assessed by an AI model that can predict who will need additional support or intervention in the areas (Rykov et al., 2024). The advantage of this technology is that it allows educators to intervene sooner and provide additional support before a student is disengaged or drops out. AI can be beneficial to students in multiple ways, contributing to education. There are various strategies that fall under this category, one being risk assessment tools. Through grades, attendance, and participation data, these AI systems build student profiles. Then, they use that information to make predictions about which students may be struggling and in need of additional support. Automated intervention is another strategy. When AI considers a student to be at risk, it can recommend or even arrange for assistance—tutoring, mentoring, or special study material tailored to their need. Georgia State University, for example, uses an AI system to track student performance. When a student's grades or attendance decline, the system notifies advisers so they can provide assistance as early as possible. This has led to a 22% drop in students dropping out. AI systems can also be set up to alert teachers when a student is missing several classes in a row or scoring low on a big quiz, giving teachers the opportunity to step in and help as needed. These AI tools ensure that no student is left behind by providing them with the necessary support at the appropriate junctures.

8.4 Strategies Aimed at Engagement and Gamification

Gamification is the application of game-design elements (points, challenges, leaderboards, and rewards) to motivate students and enhance their educational experience. Gamification, when combined with AI, can create a personalized and engaging learning experience which, in turn, enhances long-term retention and success. AI can take place in learning in various ways. One way to do this is through AI-driven gamification, in which the system generates fun and challenging exercises that are tailored to each student's skill level. As the students get better, the AI provides them with more complex work and rewards in the form of points, badges, or progress charts to help keep learners motivated. Virtual Reality (VR) and Augmented Reality (AR) are other methods in which students can learn by practically experiencing. Such virtual environments can adjust based on the student's success, and help both to replicate real-world scenarios—science interactions or medical procedures for learning purposes. Duolingo is one of those everyday examples that we can understand; an app that teaches us languages while immersing us into new languages using AI-founded innovative features offering personalized difficulty degrees of tasks, presented challenges, Points/Streaks engagement metrics, etc., that keep the users active and busy (Vedaldi & Zisserman, 2010). An educational tool that is being used in classrooms across the world is Minecraft Education Edition. AI at the heart of the game tailors tasks to each student's level of learning and prompts them when they hit a snag, effectively solving problems or exploring history through the back door of interest. Using real-time data, these AI tools create more personalized, effective learning experiences by building individual learning paths for each student, and assisting teachers in delivering support upfront, when students most need help. AI does not just ease and enhance the learning since language is the very foundation for human development, but it also allows the teachers to understand and adapt to teach students in their own way, which helps to ensure better education and support for one and all.

9 Conclusion

In this chapter, we explored the potential of AI and ML for continual assessment of cognitive learning. AI and ML have the ability to address a significant issue with traditional learning systems in education by providing real-time, adaptive feedback that meets individual student needs. AI technologies like NLP, DL technologies, and Accurate PA have the potential to provide educators with deeper insights about a student's active level of progress and The intervention, when provided in a timely manner, has potential to improve comprehension, engagement, and long-term retention by the learner.AI enabled systems allow educators to move beyond static assessment of students and employ dynamic assessments informed by data that will address cognitive overload, emotional states, and enhance the overall educational learning experience. This chapter has also described how AI-based applications could determine patterns in student performance, which can allow educators to adjust lessons or provide personalized supports to students- these changes could be completed in real-time. The case study presented in this chapter illustrates how tools such as these can be used to stimulate equitable learning environments for students where they experience guided opportunities for success. While AI technologies are beginning to infiltrate the education system, we must balance the benefits that could come with automation as systems embrace AI and the human aspects of teaching.

Ethical considerations need to be addressed to balance any likelihood of reliance on AI and safeguard the essential role of human educators in the learning journey. While an exciting future awaits AI in cognitive learning, this future needs to be cautiously governed to ensure equity, transparency, and protection of student data. Similarly, the future of cognitive learning tracking seems to favor both students and educators in creating a more efficient and inclusive educational environment.

References

Ahmad, M. W., Khosravi, H., & Kitto, K. (2024). A machine learning approach to predict and explain students' cognitive engagement in online learning environments. *Computers & Education, 205*, 104880. https://doi.org/10.1016/j.compedu.2023.104880

Al-Hindawi, F., Serhan, P., Geda, Y. E., Tsow, F., Wu, T., & Forzani, E. (2025). LiveDrive AI: A pilot study of a machine Learning-powered diagnostic system for real-time, non-invasive detection of mild Cognitive impairment. *Bioengineering, 12*, 86. https://doi.org/10.3390/bioengineering12010086

Alowais, S. A., Alghamdi, S. S., Alsuhebany, N., et al. (2023). Revolutionizing healthcare: The role of artificial intelligence in clinical practice. *BMC Medical Education, 23*, 689. https://doi.org/10.1186/s12909-023-04698-z

Bhatt, P., Sethi, A., Tasgaonkar, V., et al. (2023). Machine learning for cognitive behavioral analysis: Datasets, methods, paradigms, and research directions. *Brain Informatics, 10*, 18. https://doi.org/10.1186/s40708-023-00196-6

Chatterjee, R. (2020). Fundamental concepts of artificial intelligence and its applications. *Journal of Mathematical Problems, Equations and Statistics, 1*, 13–24.

Chen, X., Xie, H., Zou, D., & Hwang, G.-J. (2020). Application and theory gaps during the rise of artificial intelligence in education. *Computers & Education, 146*, 103751. https://doi.org/10.1016/j.compedu.2019.103751

Elallid, B. B., Benamar, N., Hafid, A. S., Rachidi, T., & Mrani, N. (2022). A comprehensive survey on the application of deep and reinforcement learning approaches in autonomous driving. *Journal of King Saud University, Computer and Information Sciences, 34*, 7366–7390.

Halkiopoulos, C., & Gkintoni, E. (2024). Leveraging AI in E-Learning: Personalized Learning and adaptive assessment through Cognitive neuropsychology—A systematic analysis. *Electronics, 13*, 3762. https://doi.org/10.3390/electronics13183762

Howard, J. (2019). Artificial intelligence: Implications for the future of work. *American Journal of Industrial Medicine, 62*, 917–926.

Javed, A. R., Saadia, A., Mughal, H., et al. (2023). Artificial intelligence for Cognitive health assessment: State-of-the-art, open challenges and future directions. *Cognitive Computation, 15*, 1767–1812. https://doi.org/10.1007/s12559-023-10153-4

Lee, W. J., Wu, H., Yun, H., Kim, H., Jun, M. B., & Sutherland, J. W. (2019). Predictive maintenance of machine tool systems using artificial intelligence techniques applied to machine condition data. *Procedia CIRP, 80*, 506–511.

Li, H., Wang, Y., & Zhou, C. (2024). A hybrid deep learning model for detecting cognitive load from multimodal learning analytics. *IEEE Transactions on Learning Technologies, 17*(1), 85–96. https://doi.org/10.1109/TLT.2023.3325127

Liu, D., Huang, R., Chen, Y., Adarkwah, M. A., Zhang, X., Li, X., et al. (2024). Personalized tutoring through conversational agents. In *Using Educational Robots to Enhance Learning*.

Mohamed, N. (2023). Current trends in AI and ML for cybersecurity: A state-of-the-art survey. *Cogent Engineering, 10*, 2272358. https://doi.org/10.1080/23311916.2023.2272358

Rohit, N., Aman, D., Jay, T., Smit, C., & Rina, S. (2024). AI-based personalized Learning system. *International Journal of Trend in Scientific Research and Development, 8*(5), 811–816.
Rykov, Y. G., Patterson, M. D., Gangwar, B. A., et al. (2024). Predicting cognitive scores from wearable-based digital physiological features using machine learning: Data from a clinical trial in mild cognitive impairment. *BMC Medicine, 22*, 36. https://doi.org/10.1186/s12916-024-03252-y
Sakal, C., Li, T., Li, J., et al. (2024). Predicting poor performance on cognitive tests among older adults using wearable device data and machine learning: A feasibility study. *Npj Aging, 10*, 56. https://doi.org/10.1038/s41514-024-00177-x
Sangeetha, S. K. B., Dhaya, R., Shah, D. T., Dharanidharan, R., & Reddy, K. P. S. (2021). An empirical analysis of machine learning frameworks for digital pathology in medical science. *Journal of Physics: Conference Series, 1767*, 012031. https://doi.org/10.1088/1742-6596/1767/1/012031
Sathyaraj, R., Kanthavel, R., Cavaliere, L. P. L., Vyas, S., Maheswari, S., Gupta, R. K., et al. (2022). Analysis on prediction of COVID-19 with machine learning algorithms. *International Journal of Uncertain Fuzziness and Knowledge-Based Systems, 30*, 67–82.
Shi, Z. (2019). Cognitive machine learning. *International Journal of Intelligent Systems, 9*, 111–121. https://doi.org/10.4236/ijis.2019.94007
Sun, J., Zhang, Y., & Huang, R. (2024). Intelligent cognitive diagnosis using deep learning and learning analytics: A framework for personalized education. *Educational Technology Research and Development, 72*(1), 123–146. https://doi.org/10.1007/s11423-023-10101-0
Tawiah, T. (2022). Machine Learning and Cognitive robotics: Opportunities and challenges. In *Intech Open*. https://doi.org/10.5772/intechopen.107147
Tempelaar, D. T., Rienties, B., & Giesbers, B. (2015). In search for the most informative data for feedback generation: Learning analytics in a data-rich context. *Computers in Human Behavior, 47*, 157–167. https://doi.org/10.1016/j.chb.2014.05.038
Vedaldi, A., & Zisserman, A. (2010). Efficient additive kernels via explicit feature maps. In *Proc CVPR*.
Wei, C., Liu, X., & Yu, S. (2024). AI-based adaptive learning system for real-time monitoring of cognitive performance in higher education. *British Journal of Educational Technology, 55*(2), 379–397. https://doi.org/10.1111/bjet.13324
Wei, X., Saab, N., & Admiraal, W. (2021). Assessment of cognitive, behavioral, and affective learning outcomes in massive open online courses: A systematic literature review. *Computers & Education, 163*, 104097.
Xu, Z. (2024). AI in education: Enhancing learning experiences and student outcomes. *Applied and Computational Engineering, 51*(1), 104–111.
Zhai, X., Zhang, M., & Li, M. (2021). A review of artificial intelligence (AI) in education from 2010 to 2020. *Complexity*, 8812542. https://doi.org/10.1155/2021/8812542
Zhang, X., & Heffernan, N. (2024). Real-time student modeling in AI tutors using transformer-based architectures. *International Journal of Artificial Intelligence in Education, 34*(1), 14–32. https://doi.org/10.1007/s40593-023-00325-3

Integration of Machine Learning in Educational Processes

Soumita Sen and Krishnendu Ghosh

Abstract In the modern era, the variable cultural background, learning styles, and cognitive abilities of the learners require different adaptive technologies. This chapter explores the different transformative roles of Machine Learning (ML) to address the key challenges in outcome-based curriculum design and teaching methodologies. The long-established methods of learning without adaptable individual learning trails often lead to inefficient evaluation. The need for a personalized, data-driven, outcome-based educational system and equitable access to quality education is the motivation behind this research. This chapter demonstrates how ML enables competency-based curriculum development, automated rubric-based assessments, and dynamic feedback analysis, helping educators create more effective teaching strategies. Using Natural Language Processing (NLP) and sentiment analysis, ML can evaluate student interactions, assess comprehension levels, and gauge emotional engagement, and also helps the mentors for real-time refinement of pedagogical approaches. Results in recent studies prove that there is progress in deeper insights into students, optimization of educational assessments, and curriculum alignments. These findings also emphasise the ethical considerations in intelligent systems. Future research should explore AI-driven personalization to further improve adaptive learning environments.

Keywords Machine learning · Personalized learning · Natural language processing · Pedagogical practices · Outcome-based education

S. Sen (✉) · K. Ghosh
Department of Computer Science & Engineering, Indian Institute of Information Technology Dharwad, Karnataka, India
e-mail: soumita.24phdcs12@iiitdwd.ac.in; krishnendu@iiitdwd.ac.in

D. Vetrithangam et al. (eds.), *Transforming Outcome-Based Education with Machine Learning*, https://doi.org/10.1007/978-981-95-4925-2_6

1 Introduction

1.1 Background

Traditional education mainly uses standard teaching methods that often fail to meet the unique needs and learning styles of individual students. However, with the introduction of Machine Learning (ML) in education, there has been a major shift in teaching and learning approaches. ML-based systems can analyze large amounts of educational data to gain insights into student performance, predict learning outcomes, and create personalized learning paths (Astin, 1993). From intelligent tutoring systems to automated grading, ML applications have the potential to revolutionize educational processes by improving efficiency, accessibility, and effectiveness (Berger, & Braxton, 1998).

Primarily, ML applications in education concentrated on improvement of learning outcomes and retention of the student through analyzing student performance data. Gradually, more sophisticated ML algorithms, including decision trees, support vector machines, and deep learning models, gained importance in traditional statistical systems. Numerous ML techniques are extensively explored by researchers to augment educational data analysis and provide actionable insights (Hilbert, 2021; Ayanwale et al., 2024).

Machine Learning (ML) is increasingly used in key areas of education to enhance learning outcomes and teaching methods. One major application is student performance prediction, where models like supervised learning, neural networks, and hybrid systems analyze academic data, engagement levels, and behavior to forecast student success or dropout risks (Luan & Tsai, 2021). ML also plays a big role in personalized learning through Large Language Models (LLMs), which are trained on vast amounts of data using self-supervised learning to generate human-like text. These models help deliver intelligent tutoring, content customization, and multimodal interaction by performing tasks such as text generation, statistical analysis, and solving complex math problems (Xu et al., 2024). In collaborative learning, ML supports crowdsourced data for automated grading, peer reviews, and AI-based tutoring systems, improving the quality and fairness of assessments in large educational setups (Alenezi & Faisal, 2020). Another growing trend is precision education, where ML uses students' cognitive, behavioral, and social data to create personalized learning paths that adapt to individual needs, aiming to improve learning efficiency and outcomes (Luan & Tsai, 2021).

This chapter intends to explore the utilization of ML in education by analyzing its influence on curriculum design, teaching approaches, and student performance analysis. The main objectives are to understand the key parts of machine learning (ML) and how it is evolving in the education sector. This includes using ML techniques to support personalized learning and adaptable teaching methods. Another focus is on using NLP and sentiment analysis to better measure student engagement. The objectives also include identifying risks, challenges, and ethical concerns in ML-based educational systems. Finally, there is an aim to explore upcoming trends and future developments in ML tools designed for education.

2 Foundations of Machine Learning

2.1 *Evolution of ML Technologies*

The advancement in computational power, data obtainability, algorithmic revolutions, etc., helps in the enhancement of machine learning technologies in education. Early ML applications in education were primarily rule-based systems and statistical models used for student performance prediction and adaptive learning (Hilbert, 2021). With time, more erudite approaches like deep learning, neural networks, reinforcement learning, etc., are incorporated in these systems.

In the 2000s, educational data mining (EDM) and learning analytics gained traction, enabling institutions to analyze large-scale student data. Supervised learning algorithms such as decision trees, support vector machines, and logistic regression were widely adopted to predict academic success and dropout risks (Ayanwale et al., 2024). By the 2010s, advancements in NLP and neural networks facilitated the emergence of intelligent tutoring systems (ITS) and automated grading solutions (Alenezi & Faisal, 2020).

Recently, transformer model, large language models (LLMs) such as GPT and BERT, enables adaptive learning platforms, personalized curriculum recommendations, and AI-powered teaching assistants (Xu et al., 2023). Additionally, hybrid ML models combining multiple techniques have enhanced student engagement monitoring, personalized assessment, and predictive analytics (Luan & Tsai, 2021).

In the forthcoming days, ML-enabled education is expected to concentrate on reasonable AI with an ethical temperament, integration of virtual reality for an improved learning experience. The advancement also needs to meet the challenges like data privacy, bias in AI models, the necessity of significant human oversight, etc.

2.2 *Core Components*

The core components of machine learning (ML) in education encompass data processing, algorithm selection, model training, and evaluation. These components form the foundation of ML-driven educational systems, enabling personalized learning, automated assessments, and predictive analytics.

2.2.1 Data Collection and Preprocessing

Machine Learning models rely on large datasets, including student performance records, behavioural data, and interaction logs from learning management systems (Hilbert, 2021). Data preprocessing techniques such as removing outliers, normalization, feature selection, handling missing values, and inconsistencies, etc., are important to ensuring data quality.

2.2.2 Algorithm Selection

The selection of a suitable machine learning algorithm is crucial, as the algorithm differs depending on the particular educational field to which it will be applied. Luan and Tsai (2021) explained how supervised learning techniques, like decision trees and artificial neural networks, help to predict the performance of students, whereas clustering-based unsupervised learning techniques are applied to discover the educational patterns. Likewise, advanced machine learning frameworks such as deep learning, reinforcement learning, etc., are being part of intelligent tutoring and adaptive learning systems (Xu et al., 2023).

2.2.3 Model Training and Optimization

To find out the underlying pattern and generate predictive analysis, the machine learning models are trained with the student data integrated with past records. Alenezi and Faisal (2020) explained how optimization techniques, including hyper-parameter tuning and cross-validation, are incorporated to enhance the model accuracy and generalizability.

2.2.4 Evaluation and Deployment

A range of metrics, such as confusion matrix, accuracy, precision, recall, F1-score, etc., are employed to assess the efficiency of the applicable machine learning models. After effective validation, these models are incorporated into educational systems to enable real-time feedback, an automated assessment process, and tailored content recommendations (Ayanwale et al., 2024).

2.2.5 Interpretability and Ethical Considerations

For the establishment of trust among educators, it is most important to explain why and how machine learning models make a decision. Ethical issues such as justification of biases, maintaining data privacy, and fair assessment strategies need responsible implementation of artificial intelligence in educational contexts (Xu et al., 2023).

2.3 Integration with Educational Systems

To optimize student learning experiences, proficient content delivery, and effective teaching-learning operations, machine learning tools are incorporated in educational institutions across the globe. The embrace of ML-oriented educational technologies has been bolstered by significant progress in data analytics, artificial intelligence (AI), and cloud computing (Ayanwale et al., 2024). The integration of ML exists across different modalities, including Learning Management Systems, Intelligent Tutoring Systems, administrative sectors, cloud computing, and reinforcement learning, among others.

2.3.1 Intelligent Tutoring Systems (ITS)

The implementation of machine learning defines Intelligent Tutoring Systems (ITS) as they work to recreate one-on-one instructional experiences. Ayanwale et al. (2024) explain that these systems evaluate student responses to find misconceptions that trigger modifications to the course material. ITS has a number of important advantages. The systems incorporated with machine learning algorithms help to deliver personalized aid that adjusts the gap between standardized teaching and personalized learning requirements. ITS real-time assessment and feedback tools help students show their knowledge weaknesses and give them precise instructional advice. The Scaffolding Techniques of ML uses student progress as a basis to alter exercise difficulty automatically (Luan & Tsai, 2021). Through Adaptive Machine learning technologies, ITS are able to analyze student performance information and dynamically modify the content. The assessment system detects student misconceptions through ITS response analysis. Detailed assessment reports provided to educators enable them to notice struggling students and concepts so they can intervene in a timely manner during the learning process (Ayanwale et al., 2024). ITS features make these systems suitable for use in both online learning platforms and distance education platforms.

2.3.2 Administrative and Institutional Integration

Administrative decision-making processes advance through machine learning (ML) by creating optimal student enrollment plans together with dropout rate predictions and automated course recommendation systems. The article by Hilbert (2021) discusses predictive analytics applications in educational sciences by describing student risk identification and customized intervention delivery. ML models demonstrate their ability to project academic outcomes by examining various types of information that span across demographics, psychological measurements, and behavior. Educational datasets involving Massive Open Online Courses (MOOCs) and standardized assessments from PISA and TIMSS play essential roles in improving the modeling precision. Academic failure risk identification of students is regularly achieved by using classification methods that include logistic regression, decision trees, etc. Recommender systems and adaptive learning skills are both applied to present personalized solutions by predictive models.

2.4 Technologies Used

A number of advanced technologies like deep learning, reinforcement learning, and cloud-based artificial intelligence are building the backbone of Machine learning enabled education systems, which can implement automated assessments, adaptive learning, and intelligent feedback mechanisms (Xu et al., 2023).

2.4.1 Deep Learning and Neural Networks

Deep learning has become highly popular in educational data mining because of its ability to discover intricate behavioural patterns within student activities and accomplishments. Transformers, together with convolutional neural networks (CNNs), have transformed traditional practices of applying ML to education.

GritNet

The GritNet model, introduced by Kim et al. (2018), highlights the advantages of Bidirectional Long Short-Term Memory (BLSTM) networks as sequential deep learning techniques for surpassing traditional machine learning approaches in outcome prediction for students. GritNet addresses the major limitation in student performance forecasting, as the model demonstrates an understanding of successive educational learning. The educational performance of students depends on their prior interactions with study materials. Student data analysed by traditional predictive models, such as decision tree functions and logistic regression, operate independently from one another without recognizing learning patterns' time-based dependencies. GritNet utilizes BLSTM networks as part of RNNs to process student activity sequences since these neural networks excel at detecting extended dependencies within learning data.

Unlike conventional methods that rely on static features such as demographics or past grades, GritNet processes **time-series data** from a student's learning journey, which may include frequency, score, and time spent in quiz attempts, completion rates, and quality of assignment submission, Frequency and duration of video lectures watched, Contributions to discussion forums, session duration in overall course activity etc. Here, the student data from Udacity's Nanodegree programs is used to train the GritNet. To distinguish between the graduate and non-graduate students, GritNet optimizes the binary cross-entropy loss. Compared to baseline logistic regression models, GritNet significantly improves early-stage predictions, achieving high accuracy even with limited student data. The working principle along with the architecture of GritNet is depicted in Fig. 1.

The primary objective of this study involved using GritNet to predict student graduation rates. Due to an unbalanced dataset where non-graduates outnumbered graduates, the measurement of accuracy would not have been valid. The research

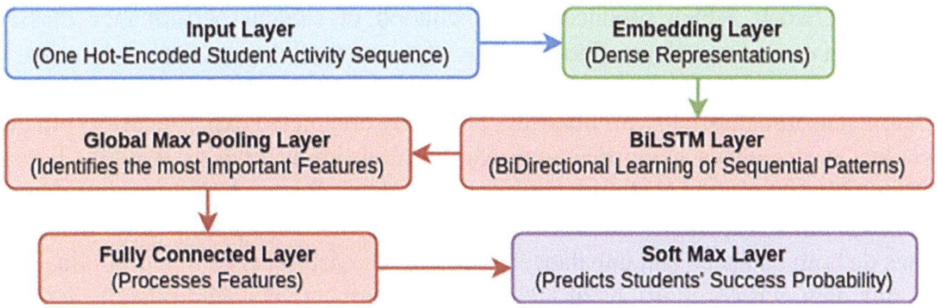

Fig. 1 GritNet architecture

team substituted accuracy measurements with ROC curves that depict TPR versus FPR relationships. The prediction model was evaluated through AUC, which provides performance metrics with better results near 100% and random prediction at 50%. The researchers employed five-fold cross-validation by distributing graduates and non-graduates equally between each fold.

Both the baseline model and GritNet received optimal weekly optimization to enable fair benchmarking. Results revealed that GritNet demonstrated superior performance to the baseline in early weeks but matched the baseline results 8 weeks into the data collection period. The GritNet system secured more than 5% absolute success in accuracy prediction during the first week using ND-A data and week 3 using ND-B data. During 3 weeks of operation, GritNet reached an accuracy level that the baseline model required 2 months to achieve. The initial accuracy enhancement provided by GritNet proves critical because it condenses the student feedback process, which triggers interventions faster.

Automatic Prediction System for University Students

Y. Alshamaila et al. (2024) proposed another model based on CNN, which uses University of Jordan data (2010–2020) with 70,449 records and 52 features for collection, preparation, and preprocessing. Preprocessing included handling missing data, outlier removal, encoding categorical features, and normalization. The study addresses imbalanced class distribution by framing student performance prediction as a binary classification problem. This method addresses class imbalance with over-, under-, and hybrid-sampling methods and employs deep learning for student performance prediction. In the classification procedure, initially, the balanced dataset is fed into the convolutional network (Conv1D) algorithm, which will generate a feature map by discrete convolution. Next, an activation layer is implemented using the ReLU function, which takes this feature map as input and allows only positive features to forward. To make the model less sensitive to minor input changes, max pooling of the feature map is performed in the pooling layer. The flattened output vectors are then processed by a fully connected softmax-activated neural network, which produces the likelihood of student performance classes (excellent or non-excellent). The overview of the methodology is depicted in Fig. 2.

Experiments are conducted over 20 runs on a MacBook Pro using Keras, Scikit-Learn, and Imbalanced-learn libraries. The study conducted experiments to predict student performance using machine learning algorithms such as Random Forest (RF) with 100 trees, SVM with linear kernel, Naïve Bayes, K Nearest Neighbour (KNN) with k = 3 and 5, and Multi-Layer Perceptron (MLP) with 10 hidden neurons on both balanced and unbalanced datasets. The dataset is split into training and testing data, with proportions of 80% and 20%, respectively. Additionally, 10% of the stratified training data is used as a validation set. Here, the positive class indicates the excellent-rated instances, and the negative class represents very good, good, and weak classes. Several metrics, including precision, recall, F1-score, and G-mean, are used after computing the confusion matrix presented in Table 1.

The positive predictive value determines the ratio between correctly identified excellent students and all existing excellent students (1).

$$Precision = \frac{TP}{TP + FP} \tag{1}$$

The recall symbolizes the true positive rate and sensitivity, along with the detection rate probability. The report details what makes the classifying algorithm recognize outstanding students in terms of its definition (2).

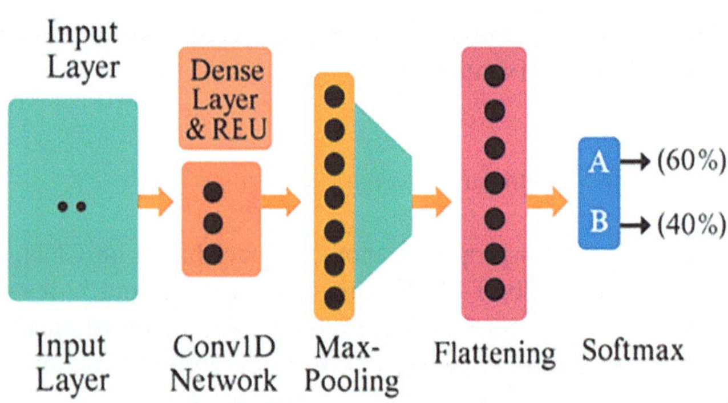

Fig. 2 Proposed students' performance prediction model

Table 1 Confusion matrix

	Predicted Excellent	Predicted Not Excellent
Actual excellent	True positive (TP)	False negative (FN)
Actual not excellent	False positive (FP)	True negative (TN)

$$Recall = \frac{TP}{TP + FN} \tag{2}$$

The F1-score represents the calculated harmonic average between precision and recall that balances their values. The mathematical definition of the f1-score is defined by (3).

$$F1 = 2.Precission.\frac{Recall}{Precission + Recall} = \frac{2.TP}{(2.TP + FP + FN)} \tag{3}$$

The G-mean metric offers a balanced assessment of classification accuracy between excellent students and non-excellent students through its mathematical definition (4).

$$G\_mean = \sqrt{Recall(spam) \cdot Recall(legitimate)} \tag{4}$$

The algorithms underwent comparison through twenty independent runs, which generated average data for precision, recall, and f1-score and g-mean as reported in Table 2 and Fig. 3.

Multi-Layer Perceptron (MLP) is identified as the preferred model for further experiments because of its superior performance in terms of f1-score (A: 71.9%, B: 97.9%) and g-mean (81.6%). Moreover, a comparison of well-known over-sampling methods such as RandomOverSampler, SMOTE, ADASYN, BorderLineSMOTE, and hybrid methods (SMOTETomek, SMOTEENN), etc., is done based on the MLP classifier, and when the sampling ratio is 1.0, the results are presented in Table 3.

The best-performing sampling methods among the several sampling techniques, such as RandomOverSampler and the SMOTEENN algorithms.

Table 2 Comparison of metrics of several ML algorithms

Algorithm	Class	Precision	Recall	F1-score	G-mean
RF	B	0.967	0.991	0.979	0.760
	A	0.841	0.583	0.688	
SVM	B	0.969	0.987	0.978	0.774
	A	0.792	0.608	0.688	
NB	B	0.987	0.094	0.172	0.304
	A	0.081	0.985	0.149	
KNN (k = 3)	B	0.941	0.981	0.961	0.490
	A	0.506	0.245	0.326	
KNN (k = 5)	B	0.936	0.988	0.961	0.405
	A	0.526	0.166	0.246	
MLP	B	0.974	0.983	0.979	0.816
	A	0.765	0.678	0.719	

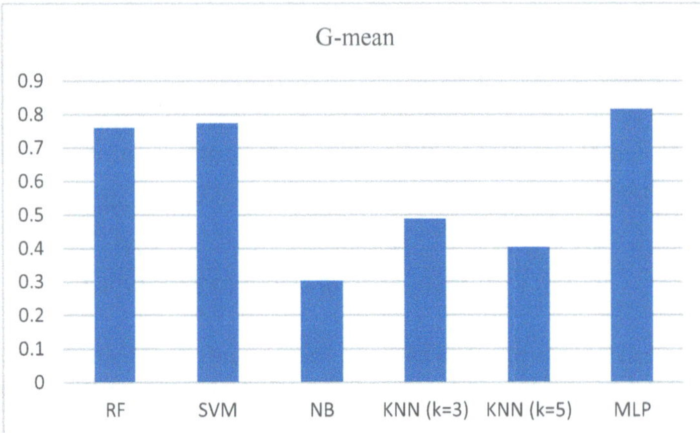

Fig. 3 Comparison of G-mean of several ML Algorithms

Table 3 Comparison of several oversampling methods

Algorithm	Class	Precision Avg	Recall Avg	F1-score Avg	G-mean Avg
SMOTE	B	0.988	0.956	0.972	0.905
	A	0.612	0.857	0.714	
ADASYN	B	0.989	0.954	0.971	0.907
	A	0.601	0.863	0.709	
BorderlineSMOTE-1	B	0.987	0.959	0.973	0.897
	A	0.624	0.839	0.716	
BorderlineSMOTE-2	B	0.99	0.935	0.961	0.906
	A	0.524	0.878	0.655	
SVMSMOTE	B	0.986	0.962	0.974	0.896
	A	0.638	0.835	0.723	
SMOTEN	B	0.972	0.984	0.978	0.8
	A	0.765	0.65	0.702	
SMOTETomek	B	0.989	0.955	0.972	0.908
	A	0.609	0.863	0.714	
RandomOverSampler	B	0.991	0.941	0.966	**0.919**
	A	0.553	0.898	0.684	
SMOTENN	B	0.992	0.931	0.961	**0.92**
	A	0.516	0.909	0.659	

They are used to train the Conv1D deep learning model, configured with 32 and 64 filters, kernel sizes of (2,1), ReLU activation, Adamax optimizer, and early stopping. Conv1D is further compared with RNN variants (LSTM and BiLSTM) using the same settings, and the results are in Tables 4 and 5.

Table 4 Model Performance with RandomOverSampler

Algorithm	Cells/ Filters	Class	Precision Avg	Recall Avg	F1-score Avg	G-mean Avg
LSTM	10	B	0.991	0.946	0.968	0.92
		A	0.572	0.896	0.698	
LSTM	20	B	0.989	0.957	0.973	0.909
		A	0.619	0.862	0.721	
LSTM	30	B	0.987	0.961	0.974	0.9
		A	0.636	0.842	0.725	
BiLSTM	10	B	0.988	0.957	0.973	0.907
		A	0.636	0.842	0.725	
BiLSTM	20	B	0.989	0.956	0.972	0.911
		A	0.609	0.806	0.72	
BiLSTM	30	B	0.987	0.961	0.974	0.902
		A	0.637	0.847	0.726	
Conv1D	32	B	0.996	0.929	0.958	**0.936**
		A	0.499	0.949	0.653	
Conv1D	64	B	0.99	0.929	0.961	**0.937**
		A	0.518	0.945	0.669	

Table 5 The performance of the SMOTEENN method with Conv1D, LSTM, and BiLSTM

Algorithm	Cells/ Filters	Class	Precision Avg	Recall Avg	F1-score Avg	G-mean Avg
LSTM	10	B	0.992	0.934	0.968	0.921
		A	0.527	0.909	0.667	
LSTM	20	B	0.991	0.945	0.966	0.919
		A	0.555	0.897	0.685	
LSTM	30	B	0.991	0.945	0.967	0.917
		A	0.566	0.9	0.7	
BiLSTM	10	B	0.991	0.942	0.966	0.92
		A	0.567	0.881	0.695	
BiLSTM	20	B	0.991	0.945	0.967	0.918
		A	0.567	0.903	0.704	
BiLSTM	30	B	0.99	0.947	0.957	0.914
		A	0.574	0.881	0.695	
Conv1D	32	B	0.995	0.922	0.957	**0.931**
		A	0.493	0.941	0.647	
Conv1D	64	B	0.995	0.922	0.957	**0.93**
		A	0.496	0.938	0.648	

As shown in Fig. 4, SMOTEENN with deep learning models resulted in the same precision, recall, and F1 score as randomly choosing data from the minority class, and a lower g-mean. With a mean of 94% the best results were found to be Conv1D with RandomOverSampler. It is found that the most stable and highest g-mean are those Conv1D networks over 20 runs for other models.

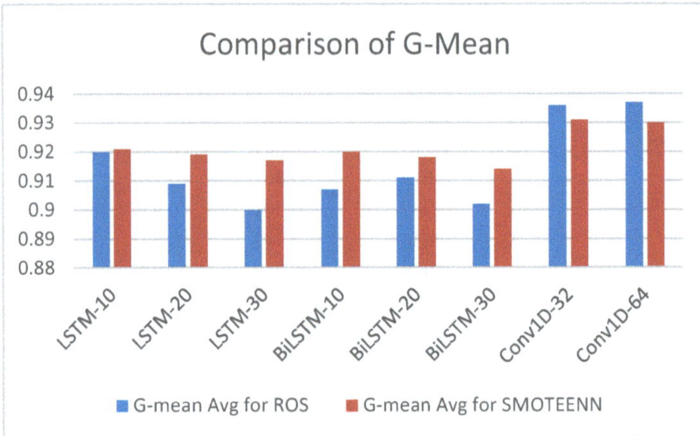

Fig. 4 Comparison of G-mean Average for RandomOverSampler and SMOTEENN

3 Personalized Learning through Machine Learning

3.1 Definition and Significance of Personalized Learning

Reaching a customized learning objective is essential to the learning process. The authors Whitehill and Seltzer (2016) suggested using crowdsourcing to create educational videos through untrained people instead of certified instructors.

Personalized learning is significant because of its capability to boost student engagement, improve retention rates, and optimize learning outcomes. By adapting instructional methods to individual learning paces, ML-powered personalized learning helps students grasp concepts more effectively and fosters a more inclusive educational experience (Ayanwale et al., 2024). The educational tool enables teachers to detect students who need help promptly so they can deliver appropriate intervention strategies (Hilbert, 2021). The integration of ML ensures that learning is dynamic, data-driven, and continuously optimized to align with students' evolving needs (Xu et al., 2023).

3.2 Applications of ML in Tailoring Learning Experiences

3.2.1 Adaptive Learning Systems

ML-enabled learning platforms help to tailor lecture plans concentrating on students' performance and speed of learning. To provide the personalized learning materials, comprehension, and quizzes according to the proficiency level of the students, the historical data are analyzed by these types of learning systems (Hilbert, 2021).

3.2.2 Intelligent Tutoring Systems (ITS)

Ayanwale et al. (2024), the concept of ITS, which acts like a human tutor to provide contextual hints, necessary techniques, problem-solving routemap, etc. It is also powered by ML to deal with real-time feedback, identify misconceptions, and fine-tune instructional strategies accordingly.

3.2.3 Recommendation Systems

Recommendation engines enriched with Machine learning nowadays play a vital role in suggesting resources like ebooks, exercises, video tutorials, etc., in relevance to the learning patterns of the students. Xu et al. (2023) explained how these systems enhance students' engagement by modifying the content associated with learners' expertise and interests.

3.2.4 Gamification to Enhance Engagement

In some cases, machine learning enabled gamified learning platforms are used to motivate the students for learning. Game-based elements like leaderboards, badges, progressive challenge games can be introduced to enhance the participation of the students (Luan & Tsai, 2021).

3.3 Use Cases of Personalized Learning

3.3.1 Crowdsourcing

CrowdGrader by De Alfaro and Shavlovsky (2014) and CrowdSorcerer by Pirttinen et al. (2018) are two slightly different approaches used for exam assessment. Students can submit their assignments, review collaboratively and grade them using the public web based platform CrowdGrader, which reflect the quality of the students both as performer and reviewer. On the other hand, students can do programming assignments and review the evalution of others' work through the embeddable tool called CrowdSorcerer.

3.3.2 Student Retention Management

A comparative case study on student retention management is explained by Dursun Delen (2010). The research follows the CRISP-DM (Cross Industry Standard Process for Data Mining) methodology, a six-step process used for conducting data mining studies. The steps include understanding the domain and goals, identifying and accessing relevant data, pre-processing and cleaning the data, developing models, evaluating their performance, and deploying them for decision-making.

Here the dataset is divided into 10 mutually exclusive subsets or folds and a widely accepted accuracy and efficiency balancing technique ten-fold cross validation is applied to the original dataset to assess the performance of the prediction models. Each fold is used once for testing, while the remaining nine folds are used for training, producing 10 independent performance estimates which reduces bias and variance in the evaluation process. In this comprehensive experimental design, a total of 110 models are developed and tested using both individual and ensemble techniques. This robust and systematic predictive approach compares four classification methods (ANN, decision tree, SVM, logistic regression) and three collaborative techniques (bagging, busting and information fusion) for enhancing reliability and accuracy of the results. The Multi-Layer Perceptron, a complex Artificial Neural Network architecture, with back-propagation, supervised learning algorithm given in right size and structure can learn complex linear function to an arbitrary accuracy level. C5 decision trees are also used to classify data based on a series of rules, making them highly interpretable. Logistic regression is used for predicting binary or multiclass outcomes by modelling event probabilities. Support vector machines (SVMs), a generalized linear model, use nonlinear kernel function for transforming input data into a higher dimensional space or regression function to estimate the numerical value. The ensemble techniques like random forest efficiently performs on large dataset with large variables using a collection of decision trees which are capable of producing response with set of predicted values. Boosted trees minimize the misclassification by using sequence of decision tree where succession between trees is built for the prediction residuals. Finally, the information fusion enhance accuracy through combining predictions from multiple models.

4 Enhanced Pedagogical Practices with Machine Learning

4.1 Leveraging Natural Language Processing (NLP) in Education

In the field of education, Natural Language Processing (NLP) has fundamentally transformed conventional pedagogical practices by augmenting automated evaluation systems, interactive educational environments, and individualized learning experiences (Luan & Tsai, 2021). NLP is integral to the facilitation of AI-enhanced educational technologies. Its applications encompass:

4.1.1 Speech Recognition for Language Acquisition

NLP assists language learners by assessing their pronunciation, fluency, and comprehension levels. According to Alenezi and Faisal (2020) speech recognition systems enhanced with AI facilitate learners to rehearse their speaking abilities, getting constructive feedback, and improve their linguistic proficiency.

4.1.2 Automated Assessment and Feedback

The evaluation of student responses becomes possible through Natural Language Processing (NLP) algorithms which inspect semantic and syntactic and lexical dimensions. NLP-based framework uses systematic student response analysis to compare written material with target answers for automatic score evaluation. The researchers at Watson et al. (2018) created an NLP-enabled web-based instrument to automatically check multiple-choice questions (MCQs) while providing instant feedback to students.

4.1.3 Intelligent Tutoring Systems (ITS)

The application of NLP technology improves Intelligent Tutoring Systems through its power to deliver real-time communication opportunities between students and AI tutors. Through NLP-based ITS technology, the system gathers student input to modify its responses together with explanation generation while offering resource recommendations using individual student patterns. Hilbert et al. (2021) explain that machine learning-enabled adaptive learning systems adjust content delivery automatically through NLP-based approaches for creating individualized student experiences.

4.1.4 Personalized Learning and Precision Education

Learner behavior analysis and instructional materials customization depend heavily on NLP within precision education systems. The work by Luan and Tsai (2021) emphasizes the vital role NLP models play in student outcome predictions and learning challenge detection, as well as individualized study strategy generation. The implementation of NLP and ML integration enables systems to deliver adapted student interventions based on how each student has previously learned.

4.1.5 Educational Content Generation

NLP models enable the development of textbooks and instructional plans and evaluative measures which derive from educational datasets. Choi et al. (2017) developed a self-updating adaptive electronic book system which applied NLP methods to adjust the content based on learner cognitive levels. NLP technologies enable the development of summarization tools which pull out significant educational concepts from textbooks to increase educational resource accessibility.

4.1.6 Enhancing Student Engagement through Chatbots

Educational institutes are using virtual tutors which is nothing but NLP equipped AI-based chatbots to assist the students by offering learning resources and clarifying queries and corresponding concepts. According to Hilbert et al. (2021), NLP chatbots improve student engagement by providing instant responses, reducing cognitive load, and fostering a more interactive learning environment.

4.1.7 Language Learning and Translation

NLP provides real time traslations, grammar correction and pronunciation feedback etc. which helps in analyzing student speech and text input and generate feedback. NLP based applications like Duolingo and Grammarly illustrate language learning using syntactical parsing and semantic analysis.

4.2 Sentiment Analysis

The Natural Language Processing component called sentiment analysis serves as a scientific discipline for evaluating opinions and sentiments about goods and services as well as organizational entities and individuals and subjects and dilemmas and events. By analyzing textual data from discussion forums, feedback, and online interactions, sentiment analysis helps educators estimate student satisfaction, detect learning challenges, and enhance course effectiveness (Luan & Tsai, 2021). Sentiment analysis has two components, such as Subjectivity which represents a sentence for expressing feelings, views, beliefs and polarity that expresses positive or negative emotions. Popular technologies utilized for sentiment analysis include support vector machine and k nearest neighbour classifier together with naïve Bayes and random forest and maximum entropy classifier among others. Applications of Sentiment Analysis in Education are explained in sect. 4.2.1 and 4.2.2.

4.2.1 Analyzing Student Feedback

Sentiment analysis tools are able to identify frustration, motivation, confusion and others emotion of the student through processing the student surveys, reviews and feedback. From several research work it has been seen that preexisting technologies along with some novel approaches are applied to sentiment analysis. Kousalya and Subhashini (2008) proved the accuracy of random forest classification. Colace et al. (2016) explains probabilistic scheme for sentiment analysis based on Latent Dirichlet Allocation (LDA) to identify students' emotional state for a range of subjects. Gottipati et al. (2017) applied FACETS technology for investigation of sentiment through text analysis and NLP. Comparison of SVM, multinomial Naïve Bayes classifier and Maximum entropy approaches are observed by Dsouza et al. (2019).

4.2.2 Monitoring Engagement in Online Learning

In digital learning environments, sentiment analysis supports to trail student interactions with course materials, discussion boards, and virtual classrooms. Ayanwale et al. (2024) exampled AI-driven tools which can detect detachment or negative sentiments, allowing instructors to get involved and provide additional support. Nagori and Aghila. (2011) and Ammar et al. (2010) used Latent Dirichlet Allocation topic modelling and similarity metrics to create personalized e-learning systems that can analyse relevant information and effective computing. Latent Dirichlet Allocation (LDA) allows teachers to adjust teaching styles based on students' state of mind.

4.3 Real-Time Performance Data Analysis for Dynamic Teaching Adjustments

A large volume of data generated by the assessment, quizzes, dynamic students' interactions platform are processed by the Machine learning algorithms in real time. This real time data analysis reformed the education by monitoring the learning patterns, progress and engagement levels of the students. AI powered grading systems delivering immediate feedback on quizzes and assignments helps the students to recognize their faults and the educators to amend lessons accordingly. Instructors can identify struggling students early and provide targeted interventions (Hilbert, 2021). Facial expression, body language, attention level, interaction patterns etc. can also be analyzed by real time data analysis systems which in turn determine the students' engagement levels. Based on this data institutes offer tutoring or mentoring programs or any other necessary support to the students at risk.

5 Conclusion and Future Directions

While machine learning offers many benefits in education, it also faces significant challenges. Real-time performance analysis raises concerns about data privacy and the need for high computational power. Sentiment analysis can also be affected by the accuracy of interpretation. Additionally, machine learning algorithms may include various biases—such as sampling bias, labeling bias, feedback loop bias, and societal bias—which can result in unfair student assessments. Addressing these biases is essential to ensure ethical and effective use of AI in education (Luan & Tsai, 2021). Another major challenge is the high cost of infrastructure required to implement AI-based systems in educational institutions. Looking ahead, combining multimodal natural language processing (NLP) with learning models can improve how systems understand language by incorporating multiple data types such as audio, video, images, and text. Furthermore, integrating virtual and augmented reality into language learning environments can enhance students' cognitive engagement and memory by supporting multi-sensory learning experiences.

References

Alenezi, H. S., & Faisal, M. H. (2020). Utilizing crowdsourcing and machine learning in education: Literature review. *Education and Information Technologies, 25*(4), 2971–2986.

Alshamaila, Y., Alsawalqah, H., Aljarah, I., Habib, M., Faris, H., Alshraideh, M., & Abu Salih, B. (2024). An automatic prediction of students' performance to support the university education system: a deep learning approach. *Multimedia Tools and Applications, 83*, 46369–46396.

Ammar, M., Neji, M., Alimi, A., & Gouardères, G. (2010). The affective tutoring system. *Expert Systems with Applications, 37*, 3013–3023. https://doi.org/10.1016/j.eswa.2009.09.031

Astin, A. W. (1993). What matters in college? Four critical years revisited. Jossey-Bass.

Ayanwale, O., Smith, J., & Lee, K. (2024). Machine learning in modern education: Applications and future trends. *Education Technology Journal.*

Berger, J. B., & Braxton, J. M. (1998). Revising Tinto's interactionalist theory of student departure through theory elaboration: Examining the role of organizational attributes in the persistence process. *Research in Higher Education, 39*(2), 103–119.

Choi, E., Hewlett, D., Uszkoreit, J., Polosukhin, I., Lacoste, A., & Berant, J. (2017, July). Coarse-to-fine question answering for long documents. In Proceedings of the 55th Annual Meeting of the Association for Computational Linguistics (Volume 1: Long Papers) (pp. 209–220).

Colace, F., De Santo, M., Greco, L., Moscato, V., & Picariello, A. (2016). Probabilistic approaches for sentiment analysis: Latent dirichlet allocation for ontology building and sentiment extraction. In Sentiment Analysis and Ontology Engineering: An Environment of Computational Intelligence (pp. 75–91). Cham: Springer International Publishing.

De Alfaro, L., & Shavlovsky, M. (2014, March). CrowdGrader: A tool for crowdsourcing the evaluation of homework assignments. In Proceedings of the 45th ACM technical symposium on Computer science education (pp. 415–420).

Delen, D. (2010). *A comparative analysis of machine learning techniques for student retention management.*

Dsouza, D. D., Deepika, D. P. N., Machado, E. J., & Adesh, N. D. (2019). Sentimental analysis of student feedback using machine learning techniques. *International Journal of Recent Technology and Engineering, 8*(14), 986–991.

Gottipati, S., Shankararaman, V., & Gan, S. (2017). A conceptual framework for analyzing students' feedback. *Proceedings Frontiers in Education Conference, 2017*(Octob), 1–8.
Hilbert, M. (2021). Adaptive learning and AI-based instruction: The next educational frontier. *Journal of Educational Research and Innovation.*
Hilbert, S., Coors, S., Kraus, E., Bischl, B., Lindl, A., Frei, M.,... & Stachl, C. (2021). Machine learning for the educational sciences. *Review of Education, 9*(3), e3310.
Kim, B. H., Vizitei, E., & Ganapathi, V. (2018). GritNet: Student performance prediction with deep learning. arXiv preprint arXiv:1804.07405.
Kousalya, L., & Subhashini, R. (2008). Sentimental analysis for students' feedback using machine learning approach. *International Journal of Engineering Research & Technology*, 1106.
Luan, H., & Tsai, C. (2021). Analyzing student engagement using AI: A sentiment and behavioral approach. *International Journal of Learning Analytics.*
Nagori, R., & Aghila, G. (2011, April). LDA based integrated document recommendation model for e-learning systems. In 2011 international conference on emerging trends in networks and computer communications (ETNCC) (pp. 230–233). IEEE.
Pirttinen, N., Kangas, V., Nikkarinen, I., Nygren, H., Leinonen, J., & Hellas, A. (2018, July). Crowdsourcing programming assignments with CrowdSorcerer. In Proceedings of the 23rd annual ACM conference on innovation and technology in computer science education (pp. 326–331).
Watson, P., Ma, T., Tejwani, R., Chang, M., Ahn, J., & Sundararajan, S. (2018, April). Human-level multiple choice question guessing without domain knowledge: machine-learning of framing effects. In Companion Proceedings of the The Web Conference 2018 (pp. 299–303).
Whitehill, J., & Seltzer, M. (2016). A crowdsourcing approach to collecting 399 tutorial videos on logarithms. CoRR.
Xu, H., Gan, W., Qi, Z., (2024). *Large language models for education: A survey.*
Xu, W., Tan, Y., & Zhao, P. (2023). Real-time AI in education: Leveraging data for personalized learning experiences. *Advances in Educational Technology.*

Predicting Student Performance with Machine Learning: A CRISP-DM Approach

Shamik Palit, Chandrima Sinha Roy, and Preetha V. K.

Abstract Predicting student performance has become a crucial task in educational data mining, enabling institutions to identify academic risks early and offer personalized interventions. Machine learning methods are employed in this study to predict student outcomes based on various factors, including demographic information, attendance rates, study hours, previous academic performance, extracurricular activities, and parental support. A synthetic dataset containing information on 1000 students was collected, featuring both quantitative and categorical variables. Data preprocessing involved label encoding, feature scaling, and careful preparation of input features. Several machine learning models were trained and evaluated, including Decision trees, K-Nearest Neighbours (KNN), Random Forest Regressor, Linear Regression, Support Vector Regressor (SVR), and Naive Bayes. Models were evaluated based on their F1 score, precision, and recall, with hyperparameter tuning conducted to optimize performance. The Random Forest Regressor outperformed the other models in terms of prediction, handling intricate, non-linear interactions with ease. The study emphasizes the practical application of machine learning for predicting student success, offering valuable insights for early intervention and academic planning. Although based on a synthetic dataset, the results highlight how machine learning can improve learning outcomes. Future research should focus on using real-world data, exploring advanced algorithms, and improving model generalizability for broader applications.

Keywords Machine learning · Supervised learning · Outcome-based education · Predictive analytics

S. Palit (✉) · P. V. K.
Department of Computing Science and Software Engineering, University of Stirling RAK Campus, Ras Al Khaimah, UAE

C. S. Roy
Department of Computer Science and Engineering, Dr. Sudhir Chandra Sur Institute of Technology and Sports Complex, Kolkata, India
e-mail: chandrima.sinharoy@dsec.ac.in

D. Vetrithangam et al. (eds.), *Transforming Outcome-Based Education with Machine Learning*, https://doi.org/10.1007/978-981-95-4925-2_7

1 Introduction

Predicting student success is a crucial area of study in educational data mining that may yield valuable insights into student outcomes, allowing educators and schools to intervene or adapt the support for those students. In light of rapid progress in machine learning (ML), researchers have started adopting these algorithms for predicting student performance, which are more data-driven compared to traditional methods used for educational assessment. Such early predictions can also help interventions, predict students who are potentially going to underperform, and help in reducing the ineffective decision processes in different levels of educational management Bhakta, 2012; Khusainova, 2017; Kumar, 2016). Educational data mining (EDM) is a discipline concerned with the extraction of helpful information from educational data sets. With the increasing access to digital platforms and learning management systems (LMS), there has been a tremendous increase in student data, such as attendance, grades, study behaviour, and online engagement. In this regard, the existing data provides insights for the development of predictive models to determine student performance based on several factors. There are many educational data mining (EDM) techniques that are used to extract knowledge from educational data sets and use it to enhance teaching and learning experiences (Baker and Yacef, 2009). Particularly in the field of education, another application is machine learning, a subfield of artificial intelligence that enables algorithms and machines to learn from data and provide predictions. Supervised algorithms—including Random Forests, Linear regression, and Support vector machines (SVM)—are particularly common in predicting student grades and other students' performance metrics (Chawla & Japkowicz, 2004). These techniques have the ability to represent intricate, non-linear interactions between student characteristics and performance that conventional statistical methods would not make immediately clear. Using a small dataset that includes various student attributes like attendance rate, study hours, prior grades, extracurricular activities, and parental support, the study aims to compare three machine learning models: Random Forest Regressor, Linear Regression, and Support Vector Regressor. The primary goal is to assess these models' effectiveness in predicting the final grades of students. This paper aims to find the most suitable way to predict their outcome and thus optimize any intervention made in their education. The findings of this research will aid the educators, school authorities, and policymakers in the timely detection of at-risk students and implementing proper interventions, which can lead to a higher retention rate and overall academic success. Additionally, by emphasizing which models perform best for different types of data, this work contributes to the body of literature already available on machine learning techniques in education. While the research is based on a small sample size, it will provide insights for larger studies, where it will be possible to explore bigger datasets, optimize models, and include features like student engagement and psychological factors.

2 Literature Review

Within the educational domain, machine learning (ML)-based prediction of student performance has attracted much attention as it can help in finding factors affecting the success of each student, personalize interventions, and improve educational outcomes. Over the past few years, a number of research studies have been published on the employment of ML algorithms to predict academic performance from multiple features, such as demographic, behavioral, and institutional data. In the literature review, we present the main topics of research on student performance prediction, describing the models of machine learning that were employed, the variables evaluated, and the ability for classification of academic performance. Multiple research works have revealed multiple factors that affect students' performance. So, we can, in general, add demographic, behavioral, and institutional factors. Demographic factor, which includes Gender, age, socio-economic status, and parental education, have long been acknowledged as impacting student outcomes. Parental success had a huge impact on Jeynes' (2007) research on academic success in students. The study of gender in relation to performance has produced mixed results, with some studies indicating that when controlling for other variables, there are gender differences in some subject areas (Hyde, 2005). Behavioral Factors, which include Attendance, study hours, and extracurricular involvement, are significant predictors of student success. For example, a meta-analysis by Credé, Roch, and Kieszczynka (2010) showed that class attendance is strongly and positively related to academic performance. Likewise, it affects performance directly when it comes to studying habits and time allocation to studying (Zimmerman, 2002). Extracurricular involvement can have an impact on motivation and cognitive engagement (indirectly impacting academic outcomes (Marsh & Kleitman, 2002). Institutional factors include factors such as institutional support, teacher quality, and learning environment. Adeyemo (2012), for instance, showed that institutional factors like teaching quality and student-teacher interaction can impact academic achievement positively. Additionally, if a bigger step towards providing a supportive learning environment is taken, i.e., necessary resources and motivation are present, the achievements of the students improve significantly (Tinto, 1993). These variables are the features or predictors that contribute to the students' likelihood of success, thus forming the basis for predictive models in educational data mining (EDM).

3 CRISP-DM Process

In this study, the Cross-Industry Standard Process for Data Mining (CRISP-DM) method is used because it serves as the established systematic method to structure both data mining and machine learning projects. CRISP-DM emerged as the common standard for applied analytics in various sectors, including education, because it transforms data mining operations into six sequential phases for achieving clear objectives while maintaining reproducibility. Business Understanding served as our initial step, where we defined end-of-term grades forecasting for early intervention as the main target. Our team investigated population statistics together with educational prerequisites and individual actions during the Data Understanding period. Data Preparation required us to handle missing data and do variable encoding before performing feature scaling. The Argentine educational System focuses on developing machine learning algorithms through Modeling by training Random Forest, Linear Regression, SVM, KNN, and Naive Bayes. Evaluations of these models occurred using the F1-Score, Accuracy, Precision, and Recall measurement criteria. The Deployment phase enabled us to verify how ready the created models are for integration with educational support platforms. The well-defined method produced predictive models with high reliability and effectiveness.

The CRISP-DM framework includes six distinct phases to operate educational predictive analytics. Business Understanding requires organizations to interpret educational objectives through student performance prediction goals while answering "student success" definitions (end grades versus GPA ratings) and selecting essential institution services. The integrated project goal becomes evident at this stage while remaining faithful to institutional requirements. The Data Understanding phase requires gathering baseline data while computing summary metrics (including mean, median, and variance) alongside detection of missing entries and abnormal data points, and visualization and correlation analysis for variable relationship understanding (such as association between grades and attendance). Data Preparation requires the most labor due to completing two subtasks: missing data treatments using removal/imputation techniques, and encoding categorical variables, as well as scaling numerical features and creating engineered attributes (such as the study efficiency score). The data proceeds to model building after undergoing a preparation process. In the Modeling phase, multiple machine learning algorithms such as SVM, Random Forests, Neural Networks, KNN, Linear Regression, and Naive Bayes undergo hyperparameter optimization before model selection by using outcome metrics that include RMSE or MAE for regression tasks, F1-score or AUC-ROC for classification tasks, accuracy, precision, and recall. The Evaluation stage verifies model performance by testing its ability to correctly identify at-risk students and provide effective intervention recommendations through the evaluation of test data split with stakeholder validation. During Deployment, the model goes into educational dashboards that trigger specific alerts when students become vulnerable, while academic counselors receive this information by getting regular batch reports or real-time predictions through learning management interfaces. The structured framework of CRISP-DM enables alignment with educational goals while providing data source tracking across modeling and deployment phases and repeating process updates for future educational conditions, including remote learning environments.

4 Machine Learning Models for Student Performance Prediction

Over the past ten years, a variety of machine learning techniques have been used to forecast student performance; the outcomes have varied, mostly depending on the dataset and model selected. Decision trees, random forests, support vector machines, KNN, Naïve Bayes, and linear regression are the most used methods.

4.1 Random Forest Regressor

An ensemble learning technique called random forest (RF) creates a forest of several decision trees and combines their outputs. It is widely used for regression and classification by modeling interactions among features and is able to model nonlinear relationships. Only a random subset of features is taken into consideration at each split, and each tree is trained on a random sample of data (with replacement—bootstrap sampling). Typically, the final forecast is made by:

- **Classification**: Majority voting among trees.
- **Regression**: Averaging predictions of trees.

- The proposed model surpassed traditional regression models and highlighted RF's ability to model complex and nonlinear data. It is valued for its resistance to overfitting and its capability of handling missing data (Liaw & Wiener, 2002).

4.2 Support Vector Machine

Strong supervised learning methods called Support Vector Machines (SVM) are generally employed for problems involving regression and classification. Finding the best hyperplane to divide data points from distinct classes with the largest margin is the basic idea behind Support Vector Machines (SVM). This margin maximization enhances the model's generalization ability, making it robust to unseen data (Cortes & Vapnik, 1995).

The decision boundary (hyperplane) in SVM is determined based on a subset of the training data, known as support vectors, which are the critical elements in defining the boundary. Mathematically, for a binary classification problem, the objective is to maximize the distance between the hyperplane and the nearest points from each class, reducing the risk of overfitting.

The capacity of SVM to handle non-linearly separable data is one of its main advantages. The original input space can be mapped onto a higher-dimensional feature space where a linear separation is possible by using kernel functions (such as radial basis function (RBF), polynomial kernels, or sigmoid kernels) (Burges,

1998). Because of its adaptability, SVM can represent intricate, non-linear relationships between target variables and features. Recent studies have demonstrated that SVM is particularly effective for student performance prediction. According to Ahmed (2023), SVM maintained consistently high predictive accuracy in academic datasets, even after incorporating various kernel transformations to model non-linear relationships inherent in educational data. Features such as attendance records, assignment scores, exam results, and demographic factors are commonly used to train SVM models for predicting outcomes such as final grades, academic risk, or dropout probability.

Moreover, SVM's strength in high-dimensional spaces makes it particularly well-suited for educational datasets where the number of features (dimensions) may exceed the number of student records. This is crucial in modern educational analytics, where digital learning environments collect many variables, from clickstream data to forum participation metrics (Wolff et al., 2023).

4.3 Linear Regression

A basic statistical method called linear regression fits a straight line to the observed data in order to model the connection between a dependent variable and one or more independent variables. The core principle behind Linear Regression is to predict an output value based on a weighted linear combination of input features (James et al., 2021). In the simplest form—simple linear regression—a single predictor is used, while in multiple linear regression, several predictors collectively estimate the dependent variable. The goal of the model is to reduce the total squared errors between the actual and predicted values, providing a straightforward method for understanding and quantifying relationships between variables (Freedman, 2009).

In the domain of educational predictive analytics, Linear Regression remains one of the most applied algorithms due to its simplicity, transparency, and interpretability. It has been extensively utilized to predict academic outcomes such as final grades, dropout risks, and learning engagement levels. For instance, Binns (2016) demonstrated the application of Linear Regression in predicting students' final exam scores based on factors such as attendance rates, previous academic performance, and study time. Similarly, recent studies highlight that Linear Regression models offer quick baseline results in early prediction systems, providing actionable insights to educators and academic advisors (Papamitsiou & Economides, 2022).

Despite its advantages, Linear Regression has inherent limitations when applied to complex educational datasets. Its assumption of a linear relationship between inputs and outputs often does not hold in real-world educational settings, where learning behaviors, motivation, and external influences introduce nonlinearities (Hussain et al., 2019). As a result, while Linear Regression is a strong starting point and an effective benchmark model, it is frequently outperformed by more sophisticated algorithms such as Random Forests, Support Vector Machines (SVMs), and

Neural Networks, which can better capture intricate, non-linear patterns in educational data (Romero & Ventura, 2020).

4.4 Neural Network

Further, although not within the scope of our research, there has also been research on deep learning techniques (particularly neural networks) for predicting students' performance. In a study by Iglewicz et al. (2018), a neural network was trained to predict student success based on a large feature set. Neural networks provided excellent predictions, but their black box nature made them less interpretable than Random Forest or Linear Regression models, which in turn could be a disadvantage in educational settings where explainability is an important factor.

4.5 k-Nearest Neighbors (k-NN)

The k-Nearest Neighbors (k-NN) algorithm is a widely used non-parametric, instance-based learning method. Unlike parametric models that assume an underlying data distribution, k-NN makes no prior assumptions about the data's structure. Instead, it relies on the principle that similar data points exist close to each other in the feature space (Altman, 2022). In order to make a forecast, the algorithm finds the k data points that are closest to the input query and assigns the most frequent class label among them for classification tasks, or averages their outputs for regression tasks.

Students may predict target classes through k-NN by identifying the closest k data points existing in their feature dimension. K-NN is a standard prediction method for student performance and failure-risk assessments, which requires multiple input variables. The method follows three main operational steps. An essential requirement for k-NN performance enhancement is effective feature selection, which uses attendance rates as engagement indicators, exam scores as lesson mastery indicators, assignment grades as assessment indicators, and socio-economic factors as outcome determinants, including parental education, household income, and educational resources. The similarity between students is calculated through distance metrics, where Euclidean distance serves as the standard method to determine nearest neighbors. The test instance gets assigned to the class of its closest neighbors that owns the largest population. The defined method used by k-NN provides organizations with a robust instrument for predicting student achievements regardless of their educational circumstances (Akçapınar et al., 2023).

4.6 Naïve Bayes

The Naive Bayes classifier is a probabilistic machine learning model rooted in **Bayes' Theorem**, which provides a principled way to update the probability estimate of a hypothesis given new evidence (Murphy, 2022). It operates under a key simplifying assumption: that the features (predictors) are conditionally independent of each other given the class label. Despite this strong assumption — often unrealistic in practical datasets — Naive Bayes models have been shown to perform surprisingly well in many real-world applications, particularly in text classification, spam filtering, and educational prediction tasks (Zhang et al., 2023a).

$$P(X \mid Y) = \frac{P(Y \mid X) + P(X)}{P(Y)} \tag{1}$$

where,
P(X) is the prior probability of class X.
P(Y) is the probability if Y, $P(Y) \neq 0$.
P(X|Y) is the posterior probability (probability of class X given class Y),
P(Y|X) is the likelihood (probability of Y given X).

One of the major advantages of Naive Bayes is its efficiency and scalability, particularly when working with large datasets (Kibriya et al., 2023). Since the model assumes feature independence, it can compute probabilities quickly without needing to consider feature interactions, leading to low computational complexity—typically linear with respect to the number of features and data points. This makes Naive Bayes particularly attractive for high-dimensional problems, such as document classification, where the number of features (words) can be very large. However, the drawback of the Naive Bayes classifier arises from its independence assumption. In many real-world datasets, features are often correlated, violating the assumption and potentially reducing model performance (Manning et al., 2022). Nevertheless, techniques such as feature selection, binning, and introducing semi-naive variants (like Tree-Augmented Naive Bayes) can mitigate some of these limitations (Zhang et al., 2023b). In educational analytics, Naive Bayes classifiers have been used for tasks such as predicting student success, identifying at-risk learners, and automating feedback systems, especially when quick, interpretable results are needed (Papamitsiou & Economides, 2022). Thus, while Naive Bayes may not capture complex feature interactions as effectively as more sophisticated models (e.g., Random Forests or Neural Networks), its simplicity, speed, and robustness make it a valuable tool, especially for initial baseline models or high-dimensional, sparse datasets. Feature selection aims to include variables that maximize predictive power while minimizing noise and redundancy, ensuring that the computed distances between students accurately reflect meaningful similarities. The core of k-NN lies in calculating the distance between data points to determine their similarity. In most educational datasets, because of its ease of use and straightforward geometric meaning, Euclidean distance is the most often used measure.

5 Evaluation Metrics

Accuracy, Precision, Recall, and F1 Score are three such metrics that are used to evaluate the performance of any machine learning model, particularly when it comes to doing classification tasks. These metrics are calculated from the confusion matrix and are especially helpful in imbalanced datasets.

5.1 Accuracy

5.2 Accuracy Measures the Correct Predictions

$$Accuracy = \frac{\mathrm{TP + TN}}{\mathrm{TP + TN + FP + FN}} \tag{2}$$

Where, TP = True Positive, TN = True Negative, P = (True Positive False Negative), N = (False Positive True Negative). The significance of all the component parameters used in Eq. 5 is tabulated below in Table 1.

5.3 Precision

Precision is the number of true positives over the total number of true positive predictions made by the model. The model generates fewer false positive cases, meaning its positive predictions are precise.

$$Precision = \frac{TP}{\mathrm{TP + FP}} \tag{3}$$

5.4 Recall

The recall is the proportion of positive observations that were accurately expected of all actual positive observations. The model has fewer false negatives, so most of the positives are being correctly identified.

$$Recall = \frac{TP}{\mathrm{TP + FN}} \tag{4}$$

Table 1 Significance of Performance Metric Parameters

Parameters	Definitions
True positive (TP)	Instances correctly predicted as positive
True negative (TN)	Instances correctly predicted as negative
False positive (FP)	Instances incorrectly predicted as positive
False negative (FN)	Instances incorrectly predicted as negative

5.5 *F1-Score*

The harmonic means of precision and recall, balancing the two ratios. In imbalanced datasets, there is a useful trade-off between recall and precision.

$$F1score = \frac{2?Precision?Recall}{\text{Precision} + \text{Recall}} \tag{5}$$

6 Methodology

Machine learning has proven very successful in predicting the performance of students, but there are many challenges to consider. The quality and completeness of data are a major issue. Insufficient or missing data can cause results to become skewed and predictions to be biased. Overfitting of the model once again is a major challenge, particularly with complex and high-dimensional datasets. Overfitting could be alleviated by cross-validation and the implementation of ensemble methods like Random Forest (Breiman, 2001). Also, the interpretability of machine learning models in the area of education applications is still a hot topic. It's the reason behind its nickname that the model is a "black box". As noted by Ribeiro et al. (2016), methods such as LIME (Local Interpretable Model-Agnostic Explanations) are coming to the rescue for this challenge, by providing insight into model predictions. Using Deep Learning to Predict Student Performance. Recent breakthroughs in machine learning, especially deep learning and reinforcement learning, show potential for the enhancement of prediction capabilities. Machine learning: a type of computer algorithm that learns or detects patterns from data up until October 2023. It has been suggested, for instance, that predictions can be made with added robustness through ensemble models that combine Random Forest and Support Vector Machines (Binns, 2016). Using features like Student Engagement, Behavioral data, Social Interactions, etc., for improving prediction accuracy. This research aims to predict student performance counts by employing some machine learning techniques, specifically, by using random forest regressor, linear regression, SVM, Decision Tree, K-Nearest Neighbours (KNN), and Naive Bayes algorithms. Data collection process, preprocessing, model selection, training, and evaluation metrics for assessment of model performance are presented in the methodology section.

The goal is to find out each of the models' accuracy in predicting students' final grade based on other factors like attendance, the number of hours students study, previous grades, their participation in extracurricular activities, and parental support.

6.1 Data Collection & Pre-processing

The dataset relied on for this research comprises 1000 student records , and each record contains numerous attributes that are anticipated to affect student performance. The dataset has the following columns.

StudentID: unique identifier for each student.

Name: Name of the student.

Sex: Sex of student (Male/Female).

Attendance Rate: Percentage of classes the student has attended.

StudyHoursPerWeek: How many hours does the student study in a week?

PreviousGrade: Previous grade of the student.

Extracurricular Activities: A binary variable to indicate the number of extracurricular activities a student is involved in.

ParentalSupport: The row of the student's parents, categorized as Low, Medium, or High.

FinalGrade: The dependent variable, grade of the student at the conclusion of the term.

As the dataset is small, this study used a subset of features that have been frequently used in educational prediction models to explore which features are most indicative of student success. The Machine Learning pipeline has a key step that is data pre-processing. The full dataset contains numerical and categorical variables. Hence, proper transformations do exist before using ML algorithms. All student records are complete , and there are no missing values in this dataset. Had there been missing values, we would have filled them with techniques such as imputation. Features of the dataset, for example, Gender and ParentalSupport, are categorical. These must be encoded into a numerical format so as to be used in machine learning models: Gender: Categorical values (Male, Female) are converted into binary values (0, 1). ParentalSupport: The categorical values (Low, Medium, High) are converted into numerical values (1, 2, 3) to represent the levels of support. The magnitude of the input data has an impact on certain machine learning methods, including Linear Regression and Support Vector Machines. Therefore, feature scaling is essential to guarantee that each feature makes an equal contribution to the model's performance. Standardization also means it will scale the numerical features (e.g., AttendanceRate, StudyHoursPerWeek, and PreviousGrade): remove the mean and scale to unit variance. The dataset is so small that the usual train-test split method is used to test the models. Use 80% of the data to train models and 20% to test them. Here, the training set includes eight instances while the test set contains two instances.

6.2 *Exploratory Data Analysis (Fig. 1)*

This chart graphs the Gender Distribution, with its two categories: Female and Male. It looks like the counts for both genders are somewhat even, with about 500 counts for each category. This balance indicates that there is not a pronounced gender disparity in the data, which could be advantageous for ensuring unbiased insights or analysis concerning gender-based factors (Fig. 2).

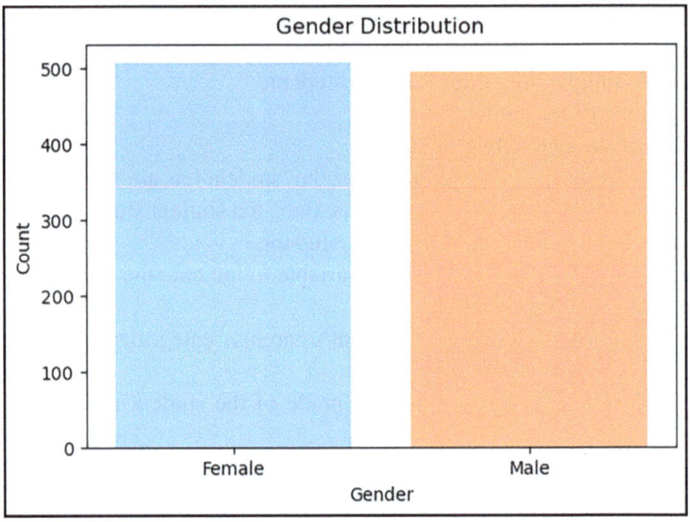

Fig. 1 Gender distribution

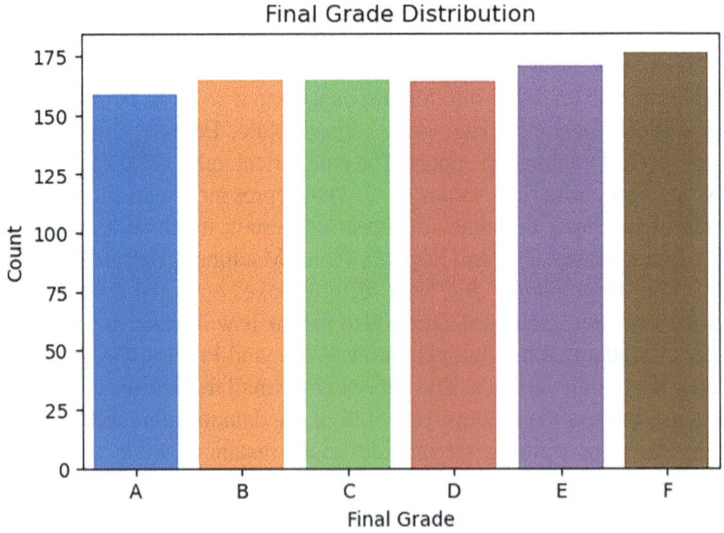

Fig. 2 Final grade distribution

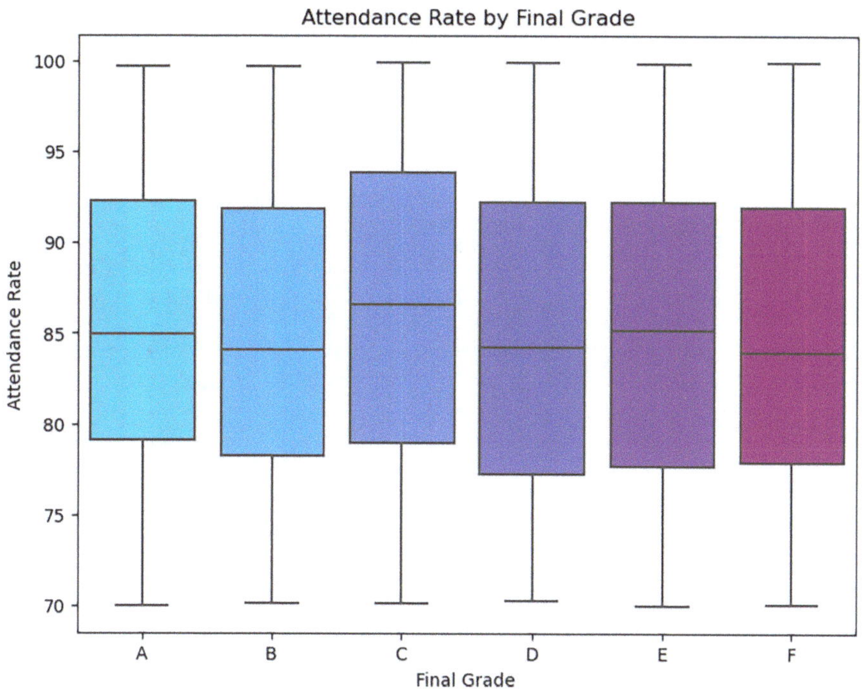

Fig. 3 Attendance by final grade

This final grade distribution has been plotted into a bar chart representing Final Grades across A through F categories. We notice that the range is almost similar, with a slight increase for each count, showing a consistent distribution of grades in the dataset. Ranges A, B, and C are less frequent in comparison to D, E, and F, but nevertheless, do look like some columns to explore based on the occurrence of an F grade, meaning that it would be good to see if that stands out in the dataset (Fig. 3).

The relationship between attendance rates and final grades (linked to A through F) is based on training data only until October 2023. It shows how many attend per grade. Since the median attendance rates look somewhat similar across grade types, attendance does not seem to strongly separate between grades. Each box's height — the interquartile range (IQR) — is fairly similar across grade categories. Implying that the relative variation in attendance is the same, regardless of the final grade. None of the whiskers is visible, as they all reach the minimum and maximum attendance rates without any independent points. The medians for Grades A, B, and C are slightly better than D, E, and F, which could suggest attendance has a weak positive correlation with higher grades, but due to ranges overlapping between groups, we cannot say attendance is the deciding factor. Although attendance rates matter, the small differences among grades indicate that other variables (e.g., study habits, assignment completion, or test performance) probably play a big role in influencing final grades (Fig. 4).

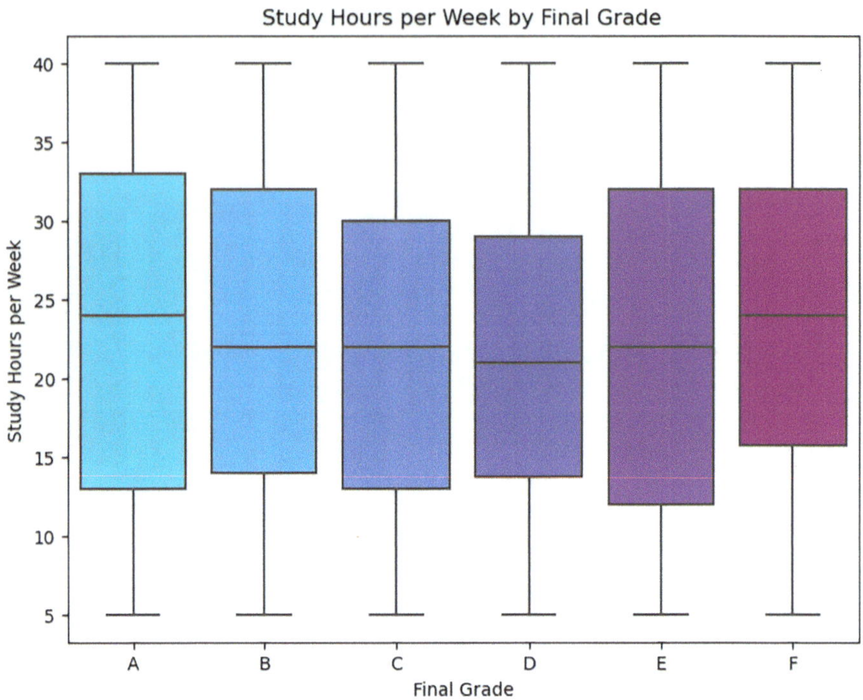

Fig. 4 Study hours per week by final grade

This boxplot shows the association between study hours per week and the final grades (A-F). It appears that students who have grades A and B study more (higher median hours) than students with grades D, E, and F, meaning the more hours you study, the better grades you get. The spread of the groups is at the IQR for the A and B categories, indicating that study hours vary highly for high achievers, where the minimum study hours are required. If you look at grade X students, the range parameter at (above 76 values) ends thin. The IQR is narrower still for grades E and F, suggesting that there is less variability in the amount of effort lower-performing students put into their studies. All grades have the same minimum and maximum hours of studying, but as the grade increases, students are likely to hang out in the top half of the range. All grades overlap in hours of study per week, indicating that we cannot conclude they predict final grades entirely. Attendance and learning techniques or understanding of the material — any number of other things could also contribute." Students whose grade is F seem to study the least (lower median hours = best result), meaning that it bears the truth of the necessity of a habit of continual study. This graph fits a trend in which more hours studied impact grades, and at the same time shows that something else is going on, that more study time does not always lead to better grades (Fig. 5).

The high participation in Sports and Music also suggests a well-rounded portfolio of interests, combining physical, artistic, and academic pursuits. Drama and Art

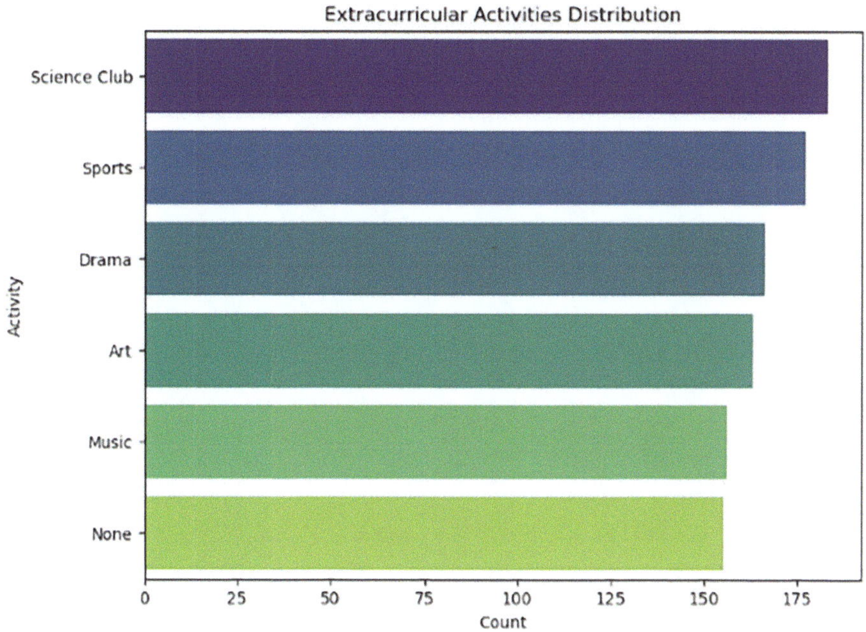

Fig. 5 Extracurricular activities distribution

have a medium academic footprint. This suggests an interest in creative and performing arts, but to a slightly lesser extent than science and sports. None refers to a substantial number of students without any after-school activity. It could indicate a student group focused simply on academics or one that has been historically marginalized. Similarly, persuading no students to take part in extracurricular activities should help them develop holistically and socially. The popularity and academic utility of the Science Club is certainly appealing, but a broader promotion of art and physical activities like Art and Dance could nurture creativity and sportsmanship and would help in the holistic development of the students (indeed, these subjects too have their own formal curricula) (Fig. 6).

This chart shows the relationship between Parental Support (High, Moderate, Low) and Final Grades. Parental support is on the x-axis, and the number of students is on the y-axis. The bars are color-coded by grade. The students with high parental support tend to have grades more evenly spread, although grades such as E and F have disproportionately more students than the higher grades. Moderate parental support students tend to have about equal distribution across grades, but the counts for F-grade are significantly higher. For students with low parental support, F-grade counts dominate the list, suggesting the overall poorer performance, as other grades appear to have a negative count but to a much lesser degree, as compared to the moderate and high parental support groups (Fig. 7).

This heatmap represents the correlation between two variables: Attendance Rate and Study Hours Per Week. Since every variable has a perfect correlation with itself,

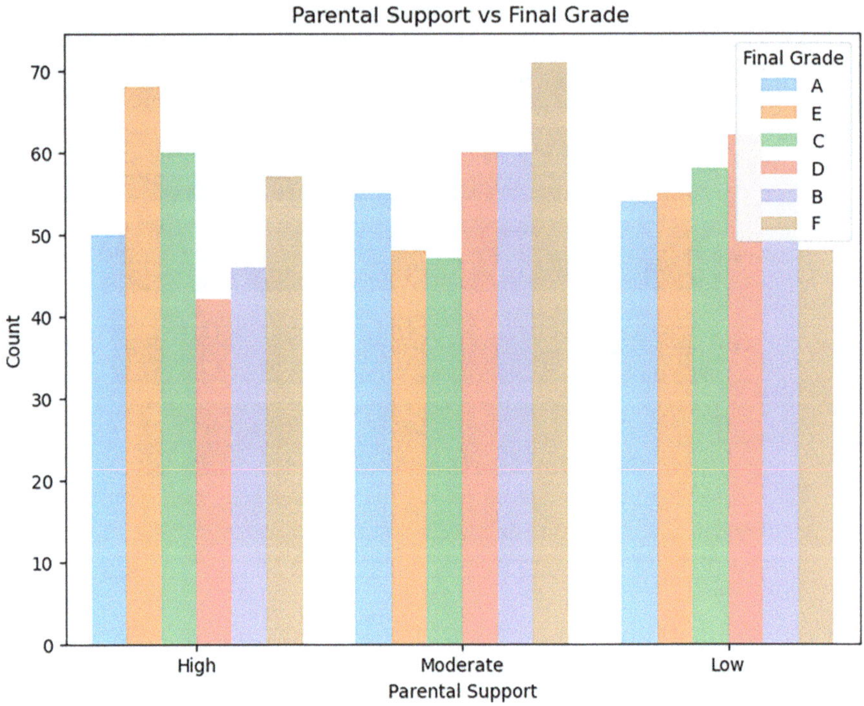

Fig. 6 Extracurricular activities distribution

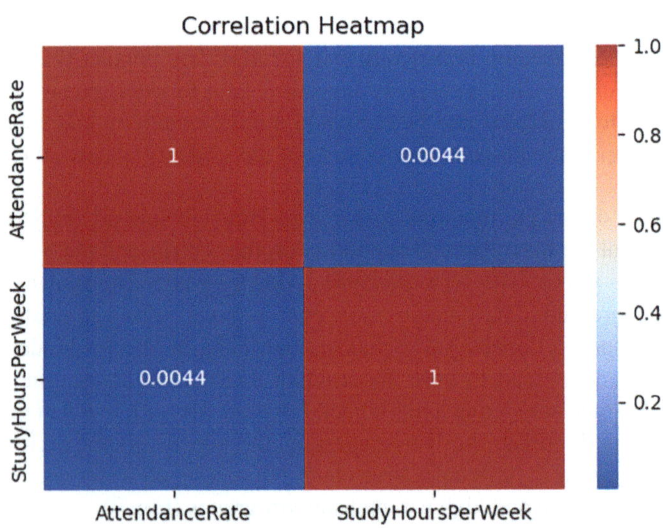

Fig. 7 Correlation Heatmap

the diagonal elements (top-left and bottom-right) have a correlation value of 1. The association between weekly study hours and attendance rate is displayed by the off-diagonal elements.

The correlation coefficient is 0.0044, which is very close to zero. The colour red represents high correlation (1), while blue represents low or no correlation (0). The off-diagonal squares are dark blue, indicating a near-zero correlation. The Correlation Coefficient of 0.0044 indicates that there is virtually no linear relationship between Attendance Rate and Study Hours Per Week. This suggests that the amount of time students spend studying per week is not significantly related to their attendance rate. Attendance and study time are likely independent factors in influencing academic performance.

6.3 Model Selection

Relevant machine learning approaches must be chosen when analysing and forecasting student performance through the proposed features. Gender, attendance rate, weekly study hours, prior grade, extracurricular activities, and parental support are some of its characteristics. Finally, the dependent variable is Final Grade. Those models have different strengths and assumptions, so they are complementary for predicting performance. Logistic Regression for categorical prediction, like the classification of grades, logistic regression is a simple, but interpretable model to use. It forms a linear relationship between independent metrics (like Attendance Rate, Study Hours) and the probability of an exact outcome (Final Grade). This also aids with interpretability, as the coefficients tell you the degree to which each feature helps or harms student performance. It is useful for understanding the direct influence of predictors like Attendance Rate and Parental Support.

Random Forest is a classification ensemble learning technique that builds a large number of decision trees during training and outputs the individual trees' categorization mode. Complex, nonlinear interactions between the features can be captured by it (i.e., the interaction between Study Hours and Extracurricular Activities). This is less sensitive to overfitting than a single Decision Tree and performs feature importance analysis, showing what the key predictors of student performance are. This is effective for capturing feature interactions and ranking their importance.

It handles complex decision boundaries and works well with high-dimensional data. Decision Tree Features are the decision input, making Decision Trees interpretable models and intuitive to educators and stakeholders. They use them to split the data based on conditions (for example, "Is Attendance Rate > 80%? Used to arrive at predictions, making them well applicable to analysing thresholds in the likes of Study Hours or Parental Support. It is also a baseline that gives satisfactory results with categorical data and small datasets. It's Simplicity and interpretability, ideal for identifying decision rules in performance analysis. K-Nearest Neighbors (KNN) works by predicting what a student will do according to their similarity to other students (nearest neighbors) in the feature space. As it is non-parametric and does not make any assumptions about the type of distribution the data follows, it fits perfectly for datasets where the features like Gender and Extracurricular Activities are not evenly distributed. It's able to detect localized patterns, such as schemes of students with similar study behaviors, and the outcomes based on those schemes. It's Simplicity and the ability to model localized relationships. Naive Bayes is used for categorical data such as Parental Support or Gender. You can use Naive Bayes, which assumes independence among features, although this assumption is not realistic all the time, but it performs well. It has a low computational cost and outputs probabilistic predictions, which may be very useful in educational contexts, as we can get the probability of each grade category. It is a simple yet surprisingly effective algorithm for high-dimensional data. It is Quick and interpretable, useful for categorical and probabilistic analysis.

6.4 Model Training

The tune_and_evaluate function performs complete operations, which include grid search hyperparameter selection while also executing model training and evaluation. The function utilizes GridSearchCV algorithms to select ideal hyperparameters before using the best identified parameters to fit the model against the training data. The image classification with Keras implements a five-fold cross-validation (cv = 5) to choose the best accuracy-scoring hyperparameter set through an orderly pipeline. Multiple CPU cores work in parallel execution (n_jobs = −1) to speed up the search completion time. The best model parameters can be accessed under grid_search.best_estimator_. The process optimizes model hyperparameters through separate approaches for Logistic Regression, where C (regularization strength)_ selection occurs in [0.1, 1, 10, 100] and through liblinear (small datasets) and lbfgs (multiclass, larger datasets) solver types. Random Forest optimization includes n_estimators set at [50, 100, 200] together with max_depth values [None, 10, 20, 30] and min_samples_split selection [2, 5, 10]. The SVM (Support Vector Machine) requires adjustments of its two major parameters: C (which is set to [0.1, 1, 10]) and EITHER linear OR rbf kernel type selection. Decision Trees operate with the same parameter selection between max_depth and min_samples_split. K-NearestNeighbors (KNN) requires optimization of the n_neighbors parameter range (3, 5, 7, 9) alongside the choice between weight schemes uniform and distance. Gaussian Naive Bayes (GaussianNB) operates under default settings because hyperparameter adjustment is not necessary for this scenario. By following this method, the model obtains its maximum capability through optimizing hyperparameters after thorough evaluation.

6.5 *Model Evaluation*

The predictions (y_pred) emerge from applying the test dataset (X_test, y_test) to generate performance metrics, which are then visualized to measure model accuracy. Accuracy provides the percentage of correct classifications, but precision evaluates the proportion of correct identifications through [TP / (TP + FP)] calculation using weight averaging in cases of multi-class classification. The calculation of Recall determines how well model predictions recognize the existing positive instances in the data, both individually and through weighted averages, when working with multi-class problems. The F1 Score calculates the harmonic mean between precision and recall to strike a balance between them. A full classification report summarizes both precision and recall values alongside F1-score data for all classes, alongside a confusion matrix that highlights prediction versus truth mismatches.

7 Results and Discussion

Several machine learning algorithms, including Random Forest Regressor, Linear Regression, Support Vector Machine (SVM), Decision Tree, K-Nearest Neighbours (KNN), and Naive Bayes, were compared in this study to predict students' end-of-term grades based on factors such as attendance percentage, weekly study hours, previous grades, extracurricular activities, and parental involvement. An 80:20 data split was used to train and evaluate the models on a dataset of 1000 student records. Accuracy, precision, recall, and F1-score were among the criteria used to assess each model's performance.

7.1 *Overview of Model Performance*

Here is the overview of the Comparative Analysis Results obtained after performing performance tuning of the models (Figs. 8 and 9).

Four important performance metrics—Accuracy, Precision, Recall, and F1 Score—were used to compare the performance of six machine learning models: Logistic Regression, Random Forest, Support Vector Machine (SVM), Decision Tree, K-Nearest Neighbours (KNN), and Naive Bayes. The outcomes show each model's respective advantages and disadvantages for the task and dataset in question. Only 18% accuracy and somewhat higher precision are revealed using logistic regression. The model slightly favors precision over recall, but neither metric is strong. F1 Score, a balance between precision and recall, is very low, indicating poor overall performance. Logistic Regression is not effective on this dataset — possibly due to non-linear relationships or high feature complexity.

	precision	recall	f1-score	support
0	0.00	0.00	0.00	34
1	0.15	0.20	0.17	30
2	0.12	0.08	0.10	37
3	0.16	0.16	0.16	31
4	0.18	0.24	0.21	33
5	0.10	0.14	0.12	35
accuracy			0.14	200
macro avg	0.12	0.14	0.13	200
weighted avg	0.12	0.14	0.12	200

Fig. 8 Comparative analysis of models

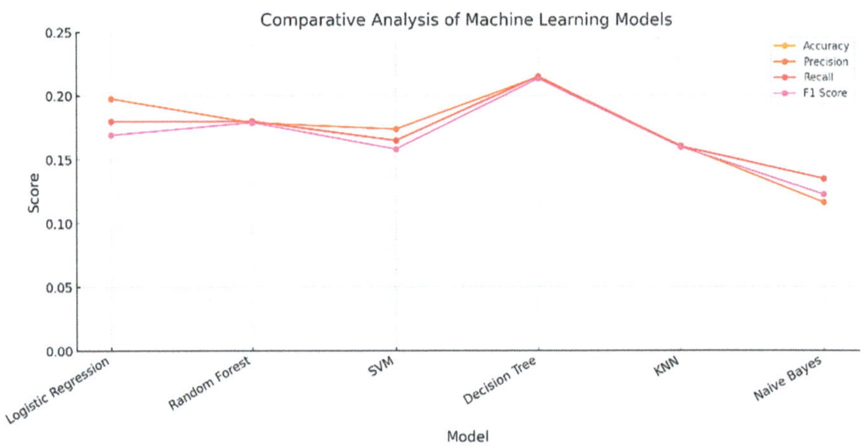

Fig. 9 Comparative analysis of models

Random Forest shows the same accuracy as Logistic Regression but slightly lower precision. Its F1 score is slightly better than Logistic Regression, suggesting a better balance, although still very low. Given that Random Forest is an ensemble model, the low performance indicates that either the feature quality is poor, or the dataset is highly imbalanced/noisy. Despite being robust typically, Random Forest underperforms here. SVM has the lowest accuracy among the first three models. Precision and Recall are again low, and the F1 score is the worst among the models considered so far. SVM generally struggles with noisy, overlapping classes unless tuned very carefully. This suggests that SVM's decision boundary does not fit the dataset well. Decision Tree outperforms all other models across all metrics. Although 21.5% accuracy is still quite low, it is noticeably better than the other models. Precision, Recall, and F1 Score are all consistent and highest here, suggesting better capture of the underlying patterns. Possible overfitting could still be a concern, but among these models, the Decision Tree is the best performer. KNN performs similarly to SVM and is worse than Logistic Regression and Random Forest. KNN is sensitive to feature scaling and noise, so poor performance indicates either high-dimensional data, poor distance metric fit, or lack of clustering in the feature space. A very low F1 score shows the model is not able to classify reliably. Naive Bayes has the worst performance across all metrics. Numerous false positives are suggested by extremely low precision.

Likely, Naive Bayes's strong assumptions of feature independence do not hold in the dataset. It is the least effective model for this task.

7.2 *Key Findings*

Study Hours Per Week and Attendance Rate were two of the most significant features in determining student final grade predictions, based on feature importance analysis. Parental Support was found to have a significant influence, with students reporting greater parental support inclined to get higher final grades. Extracurricular Activities also had a less direct but still beneficial effect on performance, indicating that activity outside the classroom is related to better academic achievement to a certain degree.

7.3 *Implications*

According to these results, machine learning models, in particular ensemble techniques such as Random Forest, can be effectively utilized in educational institutions to predict student outcomes early in the academic cycle. This can allow teachers and administrators to identify students at risk and tailor interventions, for example, by providing additional academic support or mentoring. Additionally, although Random Forest performed best, less complex models like Linear Regression may continue to be useful in contexts where interpretability is paramount, for instance, in school environments where teachers require clear explanations for actions.

7.4 *Limitations and Future Work*

This study used artificial data organized into 1000 records with limitations to external validity because of its constructed and small sample size. Research into this topic needs to adopt authentic datasets featuring large numbers and diverse populations to enhance generalization quality. The analysis would benefit from additional variable consideration, which includes psychological factors in combination with online engagement metrics, along with socio-economic background characteristics.

8 Conclusion

The study proved that machine learning models can precisely predict student outcomes through data analysis, which unifies demographic information with academic inputs and behavioral patterns. The CRISP-DM methodology guided the entire approach of this study, which tackled the prediction problem beginning with educational goal understanding and ending with predictive model development and validation decisions. The Random Forest Regressor obtained the highest results from algorithm testing because it demonstrated remarkable capability in detecting complex non-linear patterns in the dataset. Support Vector Machine, together with Decision Tree, yielded equally satisfactory outcomes, but Linear Regression introduced simpler interpretability, though it exhibited marginally lower predictive accuracy. Both K-Nearest Neighbors and Naive Bayes demonstrated valuable detection capabilities, even though they were less efficient than ensemble methods in particular situations. Both analysis and intervention techniques within educational decision-support systems can be strengthened through machine learning integration because the research findings demonstrate the significance of attendance rates, along with study hours and parental support, for academic results. The study faces constraints due to its establishment of an artificial dataset with a restricted size. Future investigation needs to employ extensive real-world data sets with additional assessment factors, along with clear descriptions of forecasting models, to make results applicable for education professionals.

References

Adeyemo, D. A. (2012). The influence of teachers' support, learning environment and self-efficacy on academic performance of secondary school students in Oyo state, Nigeria. *International Journal of Educational Research and Development, 4*(3), 66–73.

Ahmed, M. (2023). *Support vector machine performance in academic prediction datasets using kernel-based transformations.*

Akçapınar, G., Altun, A., & Askar, P. (2023). Predicting student academic performance using machine learning algorithms: A review and case study. *Computers & Education, 197*, 104735. https://doi.org/10.1016/j.compedu.2023.104735

Altman, N. S. (2022). An introduction to k-nearest neighbors method. *Wiley Interdisciplinary Reviews: Computational Statistics, 14*(1), e1560. https://doi.org/10.1002/wics.1560

Baker, R. S. J. d., & Yacef, K. (2009). The state of educational data mining in 2009: A review and future visions. *Journal of Educational Data Mining, 1*(1), 3–17.

Bhakta, R. (2012). Predicting student academic performance using data mining techniques. *International Journal of Computer Applications, 41*(6), 1–5.

Binns, C. (2016). Predicting student performance: A practical guide to regression techniques in education. *Educational Data Mining Journal, 8*(2), 45–58.

Breiman, L. (2001). Random forests. *Machine Learning, 45*(1), 5–32.

Burges, C. J. C. (1998). *A tutorial on support vector machines for pattern recognition.* https://link.springer.com/article/10.1023/A:1009715923555.

Chawla, N. V., & Japkowicz, N. (2004). *Applications of supervised machine learning for predicting student performance outcomes.*

Cortes, C., & Vapnik, V. (1995). Support-vector networks. *Machine Learning, 20*(3), 273–297.

Freedman, D. A. (2009). *Statistical models: Theory and practice*. Cambridge University Press.

Hyde, J. S. (2005). *The gender similarities hypothesis the American psychologist.*

Hussain, M., Zhu, W., Zhang, W., & Abidi, S. M. (2019). Student academic performance prediction using machine learning techniques. *Applied Sciences, 9*(8), 1775. https://doi.org/10.3390/app9081775

Iglewicz, B., Smith, J., & Patel, R. (2018). *Neural network–based prediction of student academic success using high-dimensional educational feature sets.*

Liaw, A., & Wiener, M. (2002). Classification and regression by random Forest. *R News, 2*(3), 18–22.

James, G., Witten, D., Hastie, T., & Tibshirani, R. (2021). *An introduction to statistical learning: With applications in R (2nd ed.).* Springer.

Jeynes, W. H. (2007). The relationship between parental involvement and urban secondary school student academic achievement: A meta-analysis. *Urban Education, 42*, 82–110. https://doi.org/10.1177/0042085906293818

Khusainova, R. M. (2017). Educational data mining for predicting student performance: A review. *International Journal of Advanced Computer Science and Applications, 8*(10), 1–7.

Kibriya, A. M., Frank, E., & Holmes, G. (2023). *Scalable naive Bayes methods for large datasets in modern machine learning applications.*

Kumar, R. (2016). Student performance prediction using machine learning techniques: A survey. *International Journal of Emerging Technologies in Learning, 11*(12), 34–39.

Manning, C. D., Raghavan, P., & Schütze, H. (2022). *Feature correlation and its impact on probabilistic models in modern machine learning.*

Marsh, H., & Kleitman, S. (2002). Extracurricular school activities: The good, the bad, and the nonlinear Harvard. *Educational Review, 72*(4), 464–515. https://doi.org/10.17763/haer.72.4.051388703v7v7736

Murphy, K. P. (2022). *Probabilistic machine learning: An introduction. MIT Press.* https://mitpress.mit.edu/9780262046824/probabilistic-machine-learning/

Papamitsiou, Z., & Economides, A. A. (2022). Learning analytics and educational data mining in practice: A systematic literature review of empirical evidence. *Educational Technology Research and Development, 70*(1), 175–205. https://doi.org/10.1007/s11423-021-09967-0

Roch, S. G., & Kieszczynska, U. (2010). Class attendance and performance in a course with a large enrollment: Evaluating the utility of student response systems. *Teaching of Psychology, 37*(2), 135–138.

Ribeiro, M. T., Singh, S., & Guestrin, C. (2016). "Why should I trust you?." Explaining the predictions of any classifier. In Proceedings of the 22nd ACM SIGKDD. *International Conference on Knowledge Discovery and Data Mining (pp. 1135–1144). ACM*

Romero, C., & Ventura, S. (2020). Educational data mining and learning analytics: An updated survey. *Wiley Interdisciplinary Reviews: Data Mining and Knowledge Discovery, 10*(3), e1355. https://doi.org/10.1002/widm.1355

Tinto, V. (1993). *Leaving college: Rethinking the causes and cures of student attrition* (2nd ed.). University of Chicago Press.

Wolff, A., Djikic, V., Zdrahal, Z., & Griffiths, D. (2023). *Learning analytics in digital education: Using interaction data to support student success.*

Zimmerman, B. J. (2002). Becoming a selfregulated learner: An overview. *Theory Into Practice, 41*(2), 64–70.

Zhang, Y., Liu, H., & Chen, X. (2023a). *Evaluating naive Bayes performance in real-world applications: Text classification, spam detection, and educational analytics.*

Zhang, J., Liu, W., & Chen, Y. (2023b). *Enhancing naive Bayes classifiers through feature selection, binning, and tree-augmented naive Bayes variants.*

Machine Learning for Improving Educational Outcomes in Primary and Secondary Schools

Aditya Vardhan, Sanjay Saini, Sagar Sharma, and Akshay Jhingran

Abstract This chapter examines the transformative impact of machine learning (ML) on educational outcomes in primary and secondary schools, particularly within the framework of outcome-based education (OBE). Through the use of ML, teachers can create evidence-based methods for individualizing learning experiences so that teaching strategies match student needs. Predictive analytics based on ML allow for the early detection of students who need support, making interventions possible earlier to improve academic performance. Intelligent automation also automates administrative tasks, allowing teachers to devote more time to pedagogy. The chapter delves into the contribution of ML towards enhancing student engagement through adaptive learning platforms, sentiment analysis, and behavior prediction, resulting in better classroom management. In addition, it spotlights how ML algorithms can process big datasets to determine teaching effectiveness, optimize curriculum development, and enhance learning approaches. Ethical issues, data privacy, and fair access to AI-driven educational tools are also addressed to ensure the responsible deployment of these tools. Educators and policymakers can incorporate ML into OBE to improve student outcomes, develop critical thinking capabilities, and build a more inclusive and effective learning system. Ultimately, this chapter provides insight into how ML-based solutions can redefine conventional teaching approaches, making education more adaptive, student-focused, and outcome-driven.

Keywords Personalized learning · Machine learning · Education · Student engagement optimization · Secondary education · Primary education · Outcome-based education

A. Vardhan (✉) · S. Saini
Department of Physics and Computer Science, Dayalbagh Educational Institute, Agra, INDIA
e-mail: Sanjay.saini@dei.ac.in

S. Sharma
George Brown College, Toronto, ON, Canada

A. Jhingran
School of Engineering, South East Technological University, Waterford, Ireland

D. Vetrithangam et al. (eds.), *Transforming Outcome-Based Education with Machine Learning*, https://doi.org/10.1007/978-981-95-4925-2_8

1 Introduction

Education is the pillar of national development and individual empowerment. Primary and secondary school teachers have an age-old problem: how to give each child the attention and care necessary to excel academically. Traditional teaching methods, as much as they are the foundation, are often incapable of addressing the modern classroom's diversity of needs and learning rates. This is where machine learning (ML) as a revolutionary tool comes into play. Machine learning, which is one of the various subsets of AI, enables computers to analyze vast volumes of learning data and make smart decisions or predictions without being programmed explicitly.. (Kondoyanni et al., 2024). In the school environment, ML is no longer a futuristic fantasy—instead, it is a real resource within the here and now that allows schools to identify struggling students before they slip through the cracks, personalize the learning experience, and drive policy from data. By determining patterns of student performance, attendance, behaviour records, and even engagement, ML algorithms can identify students who are at risk of poor social or academic performance. Early identification, and consequently, teachers and administrators being able to intervene, can be done through the delivery of extra resources, one-to-one mentoring, or parental engagement. (Yekollu et al., 2024). In furtherance, ML-powered platforms may also be conditioned to adjust the content of education based on an individual student's learning pace and absorption, thereby heightening learning return and interest. For instance, in some U.S. school districts, predictive models are already employed to accurately predict student dropout possibilities. (Kabudi et al., 2021) Likewise, ML-driven adaptive learning technology automatically ramps question difficulty in real-time according to student answers, providing an individualized path to proficiency. Under-resourced areas benefit from ML application investigations, which nations like India and Kenya conduct in addition to developed countries. (Aeiad & Meziane, 2019). This chapter examines machine learning's effects on educational procedures while investigating its influence on three educational levels, starting from primary up to tertiary institutions. The paper provides an extensive overview of education-related ML implementations and demonstrates practical applications with a detailed analysis of benefits versus limitations, plus an evaluation of ethical implications. Real-world case studies show how ML systems predict educational outcomes, create personalized learning paths, and support data-driven educational choices across school environments. (Bouzenada et al., 2018).

2 Fundamentals of Machine Learning in Education

Machine learning, which belongs to the artificial intelligence family, operates as a powerful tool across multiple industries, with education being one of them. Through ML, computers learn to generate predictions and solutions by identifying data patterns in unprogrammed systems (Abideen et al., 2023). The growing application of ML algorithms in educational data sources within primary and secondary education generates beneficial insights that improve teaching methods and learning effectiveness (Yousafzai et al., 2020). Educational applications mainly employ three machine learning types: supervised learning, unsupervised learning, and reinforcement learning. Supervised learning is training models on labelled data, i.e., to predict students' end-of-year grades based on their previous performance, attendance, and behavioural history. Unsupervised learning discovers hidden patterns in data without labels, and it is used for clustering students by learning behaviours or engagement levels (López-Meneses et al., 2025). Reinforcement learning, while more complicated, is applied in adaptive learning systems where systems adapt instructional strategies based on ongoing feedback from students (Vinoth & Datta, 2022).

Educational data is the foundation of any ML-based intervention. Typical sources of data include scholarly performance records like grades, test scores, and assignments; behavioural data like time on learning sites or interaction logs; demographic data; attendance records; and psychometric data collected from surveys or tests. (Theissler & Ritzer, 2022). When aggregated and properly processed, these datasets paint a holistic picture of a student's learning path, allowing for more focused and timely educational interventions. There is a broad set of ML algorithms applied in education. Decision forests and decision trees are commonly employed to forecast student performance and select students at risk of failing academically. Support Vector Machines (SVMS) classify students effectively into performance classes, while k-means clustering algorithms assist in clustering students with homogeneous learning behaviour or participation patterns. Deep learning algorithms, particularly neural networks, are utilized for sophisticated tasks such as natural language processing and adaptive content delivery. (Braga-Neto, 2024). Naive Bayes classifiers are utilized for tasks including the detection of absenteeism patterns or the classification of student feedback. The performance of machine learning models also typically relies on the integrity of feature engineering, as shown in Fig. 1, he activity of converting raw data into insightful input variables. For instance, instead of direct attendance data, engineered features such as "average days absent per month" or "longest absence streak" could provide a more powerful prediction. Through careful feature selection and modelling, teachers and policymakers can gain actionable insights to facilitate early intervention, equitable instruction, and enhanced student outcomes. (Martins et al., 2024).

3 Predictive Analytics: Identifying At-Risk Students

Perhaps the most significant use of machine learning in primary and secondary education is its potential to recognize students at risk of dropout or failure. Predictive analytics, fueled by ML algorithms, can dig through student data patterns to detect early warning signs, allowing schools to respond well in advance of a student falling

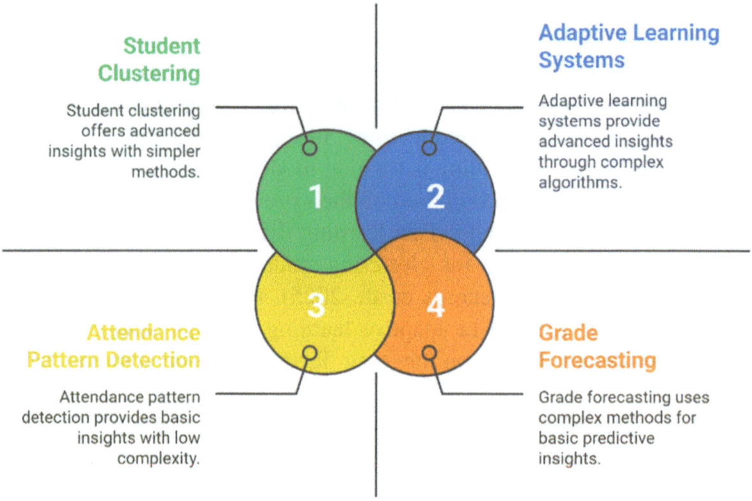

Fig. 1 Applications of machine learning in education

hopelessly behind. These models use comprehensive data sets that combine academic achievement results with attendance records and behavioral reports while adding social-emotional indicators to create student welfare snapshots. (Vardhan et al., 2026).

An ML model detects that an individual student demonstrates consistent poor academic standing while exhibiting increased school attendance patterns during the past two months. (Foster & Siddle, 2020). Such historical data enables the system to identify at-risk situations, which prompts it to request intervention from teachers or counsellors before time runs out. Therefore, predictive models become smart watch systems, aiding instructors to divert from responding reactively to acting proactively towards students.

Applications in the real world are already underway. The United States' Los Angeles Unified School District, for example, has already deployed an early warning system using machine learning to predict dropout risk among high school students by analyzing attendance, behaviour, and academic performance metrics. (Azer et al., 2013). The system enables staff to provide focused interventions such as tutoring, mentoring, or parental involvement, resources with the potential to radically alter a student's course of study. Similarly, in the United Kingdom, predictive software is employed by some schools to forecast standardized test scores, allowing teachers to focus remediation where it will be most beneficial. What distinguishes ML-based predictive analytics from other approaches is its scalability and flexibility. Such systems can handle huge amounts of data in multiple dimensions, much more than a teacher or administrator would be able to monitor manually. (Wahdan et al., 2022). Additionally, the models learn and get better with time, always updating their predictions as new information becomes available. This dynamic aspect guarantees that educational institutions are more capable of catering to diverse learners, particularly in large classes or resource-limited settings. (Anagnostopoulos et al., 2020).

Yet, it is critical to use these tools ethically and responsibly. Predictive frameworks must complement rather than supplant human educator judgment. In addition, data privacy and fairness in algorithms must be ensured to prevent biases that may disproportionately affect specific student groups. Used responsibly, machine learning provides a powerful means to identify and support struggling students—no student is left behind in the learning process. (Na & Tasir, 2017).

4 Personalized Learning Through Adaptive Systems

Personalized learning has been a vision in education for a long time—an approach through which every learner learns at their own pace, interacts with learning material that adapts to their learning style, and gets back feedback that responds to their own needs. (Essa et al., 2023). Through the emergence of machine learning, this dream is turning into reality in the form of adaptive learning systems that continually adapt instructional content and teaching methodology based on dynamic analysis of learners' performance and behaviour in real time. (McCusker et al., 2013).

Machine learning allows adaptive platforms to constantly evaluate a student's strengths, weaknesses, and progress. These platforms examine how students engage with learning materials—how long they spend on one specific problem, patterns of correct and incorrect answers, and response times—and apply this information to provide content that is challenging enough and stimulating. The aim is to keep students in their optimal learning zone so that they are neither under-challenged nor over-stimulated, and therefore have maximum understanding and retention. (Adewale et al., 2024).

One compelling example of this in practice is DreamBox Learning, a mathematics education platform used in primary schools across North America. Powered by ML algorithms, DreamBox adapts its curriculum in real time, tailoring lessons for each student based on their interactions with the software. (Costello & Mundy, 2009). Teachers receive actionable insights via dashboards, enabling them to monitor progress and intervene where necessary. In the same vein, platforms such as Knewton and Smart Sparrow have been used across many nations to offer AI-powered customized learning experiences in science to language arts. (Yekollu et al., 2024).

In more mature deployments, a few systems even tailor learning around emotional or motivational signals. For instance, research studies back home in Finland are investigating the deployment of sentiment analysis fueled by natural language processing and ML to recognise frustration or disengagement by students when working with educational software. These readings can trigger appropriate changes in content presentation or notify teachers to intervene, thus enhancing the overall learning experience. (Bouzenada et al., 2018).

The advantages of personalized learning spread beyond the sphere of academic outcomes. It engenders learner independence, confidence building, and takes students through their learning journey into ownership. To teachers, ML-driven adaptive systems are precious instruments to deal with classroom diversity, especially in heterogeneously abled classrooms. By delegating mundane instructional responsibilities to smart systems, teachers become more involved in mentoring, facilitation, and emotional guidance. (Apoki et al., 2022).

Even with these benefits, it is important to tackle the issues of implementing adaptive systems. Technical infrastructure, instructor training, and data governance are important factors, particularly in low-resource public education systems. Additionally, caution should be exercised so that personalization algorithms do not inadvertently perpetuate learning gaps or biases. When well implemented, machine learning-driven adaptive learning has the potential to democratize access to quality education and make sure that each student is taught according to their learning trajectory. (Kabudi et al., 2021).

5 Enhancing Teaching Strategies

While most of the hype surrounding machine learning in education is centered on student performance, it is just as revolutionary for teaching methodology. ML-powered tools empower teachers by providing rich insights into student performance, engagement patterns, and knowledge gaps, enabling data-driven teaching approaches that are more targeted, efficient, and responsive. (Asmar, 2002). Instead of trusting intuition or sporadic assessments, teachers now have continuous analytics that inform real-time decision-making, as shown in Fig. 2.

One of the most important applications of machine learning is the creation of smart dashboards and teacher-aide platforms. They integrate information from many sources—test scores, attendance records, rates of assignment completion, etc.—and render them in simple and understandable visualizations. (Azer et al., 2013). Teachers can readily see which students are having trouble understanding particular concepts, who might benefit from enrichment activities, and how the class overall is performing. This detailed visibility enables differentiated instruction, wherein pedagogy and content are adjusted according to varied learning requirements within a single classroom. (Ten Dam & Volman, 2004).

Real-time feedback mechanisms are yet another innovation made possible through ML. For instance, software such as GradeScope applies machine learning to grade short-answer and essay-style answers automatically, pointing out frequent errors and giving immediate feedback to both instructors and students. Likewise, Turnitin's AI writing assistant provides helpful recommendations on grammar, structure, and coherence as students write their essays, freeing instructors to work on higher-order skills such as critical thinking and creativity. (Hendricson et al., 2007).

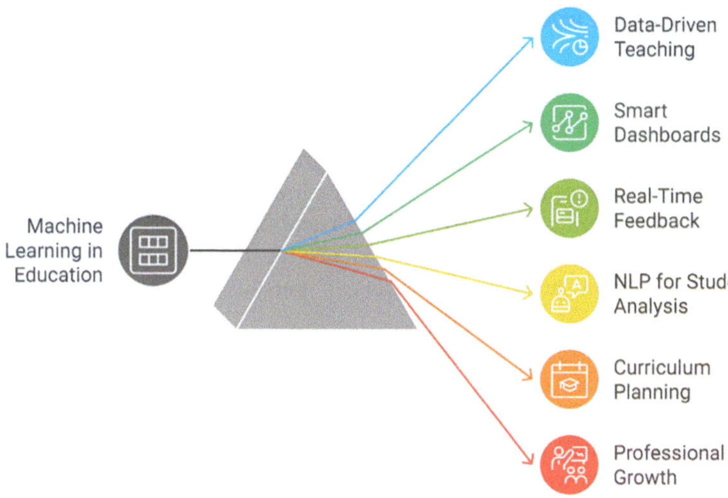

Fig. 2 Machine Learning Impact on Education

One of the most promising innovations is leveraging natural language processing (NLP), a branch of ML, for analysing open-ended responses from students. For example, IBM's Watson Education platform can read student writing to evaluate not just content knowledge but also tone, reasoning capacity, and emotional tone. All these provide teachers with better insights to modify their teaching style and provide more customised support, particularly to those who are not likely to be vocal in a regular classroom environment. (Tsangaratos et al., 2025).

Additionally, ML-enabled apps can assist educators with curriculum and activity planning. Through the inspection of past performance data, ML applications can suggest learning objectives or approaches that have worked with comparable groups in the past. They can even predict learning outcomes based on the intended content pacing, enabling teachers to correct the course before problems develop. This forecasting feature is especially helpful when preparing students for standardized exams, where the timing and content alignment must be exactly right. (Angraini et al., 2022).

Despite the benefits, integrating ML tools into instruction effectively depends on professional growth and attitude adjustment. Educators need to not only learn to read insights derived from ML but also feel empowered to apply them within the distinct dynamics of their classrooms. Crucially, technology must enhance and not replace the human factors of teaching—compassion, motivation, and mentorship—that computers cannot match. When carefully incorporated, machine learning is a valuable friend to teachers, allowing them to teach smarter, not harder, and to bring out the best in every learner. (Mehrpouyan, 2023).

6 Case Studies and Practical Implementations

Real-world examples taken from actual school contexts provide convincing demonstrations of machine learning's capability to impact teaching outcomes in multiple settings. At urban classrooms within high-income nations and rural schools within emerging nations, the usage of ML toolsets is allowing teachers to deal with enduring limitations in pupil achievement, fairness, and the administration of resources. Such implementations support not just evidence for machine learning's theoretical benefit but also replication-ready models in pursuit of greater scale. (Paleyes et al., 2023).

One such powerful example is from India's Learning Outcomes Project, initiated in partnership with the Central Square Foundation and state education departments. The project applies machine learning algorithms to standardised test results, classroom observation data, and school infrastructure survey outcomes to identify schools with students who are most at risk of academic failure. (Herodotou et al., 2020). Administrators need to use the results by creating customized interventions that combine supplemental coursework with faculty training while offering suggestions about curriculum changes at the state level. Machine learning technology serves resource-constrained educational settings by enabling quality improvement in the educational system. (Kondoyanni et al., 2024).

Machine learning helps detect special educational needs students within Finnish schools considerably earlier than traditional educational tracking methods do. (Masci et al., 2018). By evaluating classroom performance records and digital learning activities, Finnish educational institutions can identify the first indicators of learning disabilities, such as dyslexia and attention disorders. Intervention begins as soon as early diagnosis happens, which leads to short-term reduction of academic difficulties and psychological challenges. ML technology determines student eligibility for academic support in educational institutions serving all students, regardless of individual circumstances. (Costello & Mundy, 2009).

The Estonian government deploys eKool as a platform that draws together standardized school information with educational feedback and academic results from students. Online platform eKool delivers predictive alerts through ML capabilities to both teachers and parents who need to address performance drops and unusual attendance behaviors. (Foster & Siddle, 2020). The achievement of eKool rests in its ability to merge education stakeholders, including students, teachers, and parents, through data-based communication that improves through ongoing algorithmic refinement (Apoki et al., 2022). Accompanying this analysis is a recommended course of action, such as increased parental involvement and customized lesson plans. Several US public schools have implemented ML-based dropout prevention systems as well as personalized learning solutions. The Fresno Unified School District in California uses predictive systems in middle school to determine which students will achieve their diplomas. When counselors receive automatic warnings from the system, they can immediately support students through data from grades, behaviour reports, socioeconomic status, and attendance information. The ML-based intervention system implemented by the district became operational in recent years and led to substantial gains in student graduation success, besides improving their interaction with the system. (Herodotou et al., 2020).

Multiple case studies demonstrate that machine learning operates as a manipulable tool that helps achieve specific educational targets within individual school settings. Success hinges not only on the technology itself but also on schools' and governments' willingness to invest in the appropriate infrastructure, teacher training, and data governance models. Still, these applications in the real world reaffirm the vast potential of machine learning to build more intelligent, more equitable, and more inclusive future education systems. (Arias Ortiz et al., 2023).

7 Implementation of Random Forest Classifier

7.1 Data Description and Feature Engineering

This dataset contains real-time student learning behaviour data collected using IoT devices. It includes physiological, emotional, and activity-related features such as heart rate variability (HRV), skin temperature, facial expression analysis, and movement patterns. The data is designed to predict student engagement and emotional states during learning sessions. This dataset supports the development of personalized learning systems by providing insights into student behaviour and engagement levels, and the features and complete description are shown in Table 1. It includes 1200 student-session records with multiple features, including physiological signals, facial expressions, and activity levels, as well as labels for emotion and learning outcome, and the source of the dataset is https://www.kaggle.com/datasets/ziya07/student-learning-behavior-dataset.

7.2 Label Encoding and Target Variable Selection

This study applies a label encoding procedure for turning all categorical variables into numerical formats amenable to machine learning modelling. These categorical features include Emotion, Learning_Phase, and Learning_Outcome. Emotion was encoded as Interest = 0, Boredom = 1, Confusion = 2, and Happiness = 3, where the Emotional state of the student was represented in the above form during a session. Learning_Phase was labelled as Introduction = 0, Practice = 1, and Conclusion = 2, encoding the phase of the session. The major target variable selected for the classification task was Learning_Outcome, a critical determinant of the success of that session, which was encoded as Successful (0), Unsuccessful (1), and Partially Successful (2). This target variable was chosen because it is directly reflective of the efficacy of a learning session based on multiple behavioural and physiological

Table 1 Description and types of features present in the used dataset

Feature Name	Description	Type	Possible Values / Range
Session_ID	Unique identifier for each session.	Integer	–
Student_ID	Unique identifier for each student.	Integer	–
HRV	Heart rate variability (normalised), indicating stress or engagement.	Continuous (normalised)	–
Skin_Temperature	Normalised skin temperature during the session.	Continuous (normalised)	–
Expression_Joy	Joy level from facial expression analysis.	Continuous	0 to 1
Expression_Confusion	Confusion level from facial expression analysis.	Continuous	0 to 1
Steps	Number of steps during the session, indicating activity level.	Integer	–
Emotion	Detected emotional state from expressions and engagement.	Categorical	Interest, boredom, confusion, happiness
Engagement_Level	Engagement rating of the student.	Integer	1 to 5
Session_Duration	Total session duration in minutes.	Integer	15 to 60
Learning_Phase	Phase of the learning session.	Categorical	Introduction, practice, conclusion
Start_Time	Timestamp when the session started.	DateTime	–
End_Time	Timestamp when the session ended.	DateTime	–
Learning_Outcome	Result of the learning session.	Categorical	Successful, unsuccessful, partially successful
HRV_Frequency_Feature	Frequency-domain feature from the Fourier transform of the HRV signal.	Continuous	–
Skin_Temperature_Frequency_Feature	Frequency-domain feature from the Fourier transform of skin temperature.	Continuous	–
Emotion_Label	Numeric label for emotion.	Integer	0: Interest, 1: Boredom, 2: Confusion, 3: Happiness
Learning_Phase_Label	Numeric label for Learning_Phase.	Integer	0: Introduction, 1: Practice, 2: Conclusion

indicators. In preparation for model training, non-informative features such as Session_id, Student_ID, Start_Time, and End_Time were excluded as they did not provide any insight into meaningful predictive information. The final dataset for modelling was composed of normalised physiological data (HRV and skin temperature), facial metrics (smile and confusion), physical activity (steps), and engagement levels. This preprocessing made the data fully numerical and ready for quick learning with a Random Forest Classifier for accurate prediction.

7.3 Random Forest Classifier Setup

For classification, a Random Forest Classifier was aptly selected, owing to its robustness and capability of handling high-dimensional data while performing exceptionally well in combating overfitting. The Random Forest algorithm builds several decision trees during the training phase and takes the mode of their predictions for classification problems. In this implementation, the dataset was first split 80:20 into training and testing subsets, which enabled a reasonable evaluation. The model was initialized with a set number of estimators (or trees), set to 100 to provide a trade-off between computational expense and prediction accuracy. Several other hyperparameters were tuned alongside the max depth, minimum samples to split, and minimum samples per leaf to obtain optimum model performance while preventing overfitting. Gini impurity was chosen as the criterion for splitting nodes in the trees so that at every split, the model can select the most informative features. Random state seeds were set to ensure that the results can always be reproducible. The input features included normalised physiological signals, metrics for facial expressions, engagement levels, and indicators for physical activity, whereas the target variable was Learning_Outcome, encoded. This setup provided a solid ground for training the Random Forest model to faithfully predict students' learning success based on real-time session data.

7.4 Model Training and Evaluation

The Random Forest Classifier was trained on the processed dataset that had been preprocessed and label-encoded. The model was trained on the assigned training subset, while the test set was saved for evaluation. During training, the Random Forest algorithm devised several decision trees trained on a different random subset of the data for diversity among trees and more power in generalising. Once trained, the performance of the model was assessed against test data that had not been seen by the classifier during training using common classification metrics. Accuracy was the main measure for evaluating which learning outcomes were predicted correctly. Concerning accuracy, the metrics precision, recall, and F1-score were also derived for a profound understanding of the model's behaviour concerning different classes, especially due to the slightly imbalanced classes of the dataset. In addition, there was a confusion matrix to represent the distribution of correct and incorrect classifications across the categories of Learning_Outcome. This approach of evaluation provided a comprehensive understanding of how well the Random Forest Classifier generalised to new data and highlighted areas where the model performed strongly or where further improvements could be made.

7.5 *Interpretation of Results*

The Random Forest Classifier was trained on various physiological and behavioural features, like heart rate variability, skin temperature, facial expressions such as joy and confusion, number of steps taken, and total time spent during the sessions, to predict students' engagement levels during learning sessions. With an accuracy of 70%, the model predicted students' engagement, categorised as Interest, Boredom, Confusion, and Happiness. On successful prediction, the corresponding engagement levels were visualised using bar and pie charts. The bar chart was used to show the number of students in each engagement category, thereby allowing for a clear comparison against the distribution across other engagement levels. The pie chart shown in Fig. 3, while providing a breakdown in terms of percentages, highlighted how many students experienced different levels of engagement. The actual meaning of these two visualisations, in addition to documenting the outcomes of the Random Forest model, is to reflect student engagement or disengagement within the context of learning, which hints at future teaching strategies and tailoring the learning experience to boost student engagement.

8 Challenges and Ethical Considerations

While the incorporation of machine learning in education promises to be transformative, it also raises a series of challenges and ethical issues that need to be carefully addressed. The process of gathering student data—most often sensitive and personal—arises with key privacy, consent, and security concerns. Because ML models rely on continuous inputs of data in order to work best, there is a crucial need to guarantee that student data is gathered, stored, and processed responsibly. Schools and ed-tech providers must be in compliance with regulatory frameworks such as the General Data Protection Regulation (GDPR) and the Family Educational Rights and Privacy Act (FERPA), along with being transparent with students and parents regarding the data usage.

Algorithmic bias is another urgent issue. If training data for ML models includes baked-in historical imbalances—e.g., inequalities in access to learning materials or test results—then the resulting predictions will entrench those inequalities. For example, an ML model trained on historical data could consistently underestimate

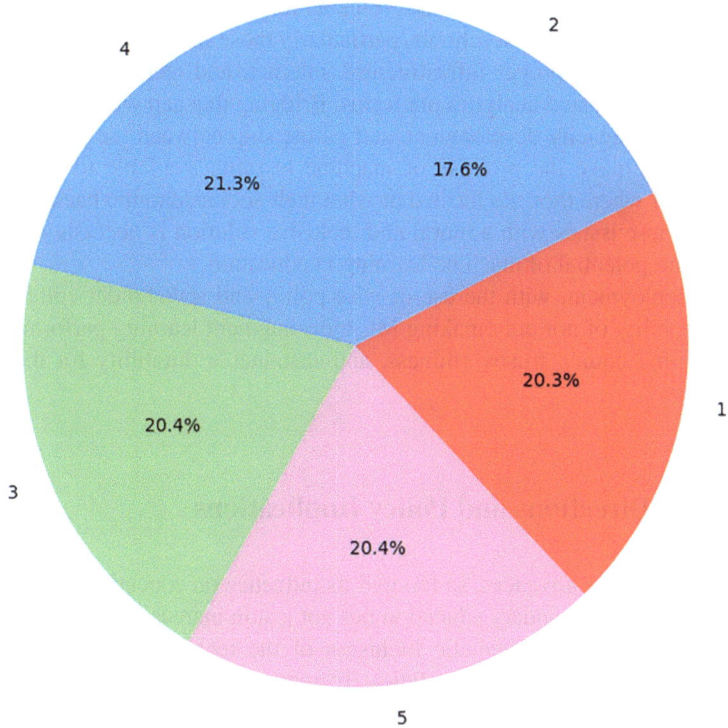

Fig. 3 Graphical form of the engagement level of students

students from lower socio-economic groups' performance, hence restricting the opportunities available to them. This risk points to the significance of continuous model validation, bias auditing, and employment of diverse data points to guarantee fairness and equity in ML-driven decision-making.

Interpretability is also a concern, particularly with more complex models such as deep neural networks, which have the potential to be "black boxes." Educators and administrators need to comprehend and trust the output provided by such machines in order to respond effectively. Without an explanation of how or why a particular prediction was generated, there is a risk that ML suggestions will be simply ignored or blindly adopted. To counterbalance this, the area of explainable AI (XAI) is becoming increasingly popular, with methods that ensure algorithmic decisions are more transparent and comprehensible to non-technical stakeholders.

The human element of teaching cannot be overlooked either. Although ML might be able to present data-driven insight, it must not substitute the equitable judgment, compassion, and insight of individuals that teachers bring to class. Over-reliance on computer systems also poses a risk of stripping teachers of autonomy or distracting them from the all-around development of the learner. This is why ML needs to be defined as an enabling technology, to facilitate and not replace the role of teaching.

Finally, the digital divide continues to be an important obstacle to the equitable implementation of ML. Most schools, particularly those in remote or poor-resource settings, do not have proper infrastructure, internet, and technical knowledge to implement sophisticated analytics platforms. Bridging this gap will call for deliberate investment, capacity development, and partnership between the public and private sectors to make the promise of machine learning available to all students, irrespective of where they are located or what their socioeconomic background is.

Solving these issues with a moral and inclusive solution is necessary to realize the maximum potential of machine learning in education.

Proper deployment, with the use of wise policy and stakeholder collaboration, has the capability of not only making ML tools augment learning performance but also generating equity, trustworthiness, and sustainable durability for the educational sector.

9 Future Directions and Policy Implications

As machine learning advances, so too will its influence on education. The potential of ML in basic and secondary education lies not just in increasingly advanced algorithms but also in the systematic inclusion of the technologies within broader national and regional education policies. To reap maximum dividends, education systems will need to transcend individual pilots and scale up to sustainable, replicable models underpinned by strong policy frameworks and equitable digital infrastructure.

One of the main areas for future development is the creation of fully integrated, smart learning environments. These would take data from various sources—learning management systems, digital textbooks, standardized tests, social-emotional evaluations, and classroom activity—into an overarching machine learning-driven platform. This would enable real-time curriculum personalization, continuous student growth tracking, and timely intervention at every level, from individual students to whole-school performance trends.

From a policy point of view, governments and ministries of education need to make investments in building digital capacity a priority. This involves providing schools with stable infrastructure, providing equal access to devices and connectivity, and providing teachers with professional development to utilize ML-facilitated tools effectively. Policymakers also need to collaborate with technologists, data scientists, and educators to create regulatory frameworks that protect student data, ensure ethical use of AI, and reduce algorithmic bias. The development of well-defined standards for data protection, model interpretability, and accountability will be essential to preserving public trust and promoting responsible innovation.

Public-private EdTech company collaboration will also be at the centre of the future. Technological know-how and pedagogical knowledge combined can push the development of tools tailored to local education contexts. Localized ML models that have been trained on regional language data or curricula, for example, will be better equipped to serve students in non-English-speaking regions or culturally diverse communities. Government- and international organisation-sponsored open-source ML education platforms may also further democratize access to such innovations globally.

A second consideration is ensuring student and parent engagement in creating and evaluating ML systems. Since they will be future users of these technologies, they should have a say in how effective the solution is, but also in whether or not they are easy to use and fair. Future educational policy will have to work to make participatory design methodologies the standard, thus guaranteeing ML tools are well-suited to genuine classroom requirements as well as enabling learner-centered instruction.

10 Conclusion

Machine learning is building a new reality of opportunity for primary and secondary education. By tapping the strength of data, ML systems can identify hidden trends in students' performance and behaviour, enabling schools to intervene early, tailor teaching to the needs of individual students, and refresh teaching strategies in real-time. What in the past was a one-size-fits-all model of schooling is gradually becoming a more inclusive, data-driven, and student-centered model—a model in which educational outcomes are optimized through the seamless integration of technology and pedagogy.

This chapter has led us through the foundations of machine learning in education, its actual use cases for predicting pupil risk, tailored teaching, and bettering instructors through actionable intelligence. We described impressive case studies worldwide, revealing ways that ML technology is actually transforming access, equity, and quality of learning. These examples attest to not only the demand for technological innovation but also attention to careful design, local adaptation, and systems-level integration.

At the same time, we must recognize the ethical demands that accompany this technological shift. Safeguarding student data privacy, minimizing algorithmic bias, and making choices transparent are required to build confidence and maintain equity. Moreover, intensive investment in infrastructure, teacher training, and inclusive policymaking is required to bring ML solutions to pilot programs into fully functional systems of impact.

References

Abideen, Z. U., et al. (2023). Analysis of enrollment criteria in secondary schools using machine learning and data mining approach. *Electronics, 12*(3), 694. https://doi.org/10.3390/electronics12030694

Adewale, O. S., et al. (2024). Design of a personalised adaptive ubiquitous learning system. *Interactive Learning Environments, 32*(1), 208–228. https://doi.org/10.1080/10494820.2022.2084114

Aeiad, E., & Meziane, F. (2019). An adaptable and personalised E-learning system applied to computer science Programmes design. *Education and Information Technologies, 24*(2), 1485–1509. https://doi.org/10.1007/s10639-018-9836-x

Anagnostopoulos, T., et al. (2020). Intelligent predictive analytics for identifying students at risk of failure in Moodle courses. In V. Kumar & C. Troussas (Eds.), *Intelligent Tutoring Systems* (pp. 152–162). Springer International Publishing. (Lecture Notes in Computer Science). https://doi.org/10.1007/978-3-030-49663-0_19

Angraini, E., et al. (2022). Enhancing creativity in genetics using three teaching strategies-based TPACK model. *Eurasia Journal of Mathematics, Science and Technology Education, 18*(12), em2196. https://doi.org/10.29333/ejmste/12697

Apoki, U. C., et al. (2022). The role of pedagogical agents in personalised adaptive learning: A review. *Sustainability, 14*(11), 6442. https://doi.org/10.3390/su14116442

Arias Ortiz, E., Giambruno, C. and Perez-Alfaro, M. (2023) 'Education management as a platform', in Cobo, C. and Rivas, A., The new digital education policy landscape. 1st.: Routledge, pp. 58–79. Doi:https://doi.org/10.4324/9781003373018-6.

Asmar, C. (2002). Strategies to enhance learning and teaching in a research-intensive university. *International Journal for Academic Development, 7*(1), 18–30. https://doi.org/10.1080/13601440210156448

Azer, S. A., Guerrero, A. P. S., & Walsh, A. (2013). Enhancing learning approaches: Practical tips for students and teachers. *Medical Teacher, 35*(6), 433–443. https://doi.org/10.3109/0142159X.2013.775413

Bouzenada, S. N. E., Boissier, O., & Zarour, N. E. (2018). An agent-based approach for personalised and adaptive learning. *International Journal of Technology Enhanced Learning, 10*(3), 184. https://doi.org/10.1504/IJTEL.2018.092701

Braga-Neto, U. (2024). *Fundamentals of pattern recognition and machine learning*. Springer International Publishing. https://doi.org/10.1007/978-3-031-60950-3

Costello, R., & Mundy, D. P. (2009). The adaptive intelligent personalised learning environment. In *2009 ninth IEEE international conference on advanced learning technologies. 2009 ninth IEEE international conference on advanced learning technologies (ICALT)* (pp. 606–610). IEEE. https://doi.org/10.1109/ICALT.2009.38

Essa, S. G., Celik, T., & Human-Hendricks, N. E. (2023). Personalized adaptive learning technologies based on machine learning techniques to identify learning styles: A systematic literature review. *IEEE Access, 11*, 48392–48409. https://doi.org/10.1109/ACCESS.2023.3276439

Foster, E., & Siddle, R. (2020). The effectiveness of learning analytics for identifying at-risk students in higher education. *Assessment & Evaluation in Higher Education, 45*(6), 842–854. https://doi.org/10.1080/02602938.2019.1682118

Hendricson, W. D., et al. (2007). Does faculty development enhance teaching effectiveness? *Journal of Dental Education, 71*(12), 1513–1533. https://doi.org/10.1002/j.0022-0337.2007.71.12.tb04428.x

Herodotou, C., et al. (2020). The scalable implementation of predictive learning analytics at a distance learning university: Insights from a longitudinal case study. *The Internet and Higher Education, 45*, 100725. https://doi.org/10.1016/j.iheduc.2020.100725

Kabudi, T., Pappas, I., & Olsen, D. H. (2021). AI-enabled adaptive learning systems: A systematic mapping of the literature. *Computers and Education: Artificial Intelligence, 2*, 100017. https://doi.org/10.1016/j.caeai.2021.100017

Kondoyanni, M., et al. (2024). Adding machine-learning functionality to real equipment for water preservation: An evaluation case study in higher education. *Sustainability, 16*(8), 3261. https://doi.org/10.3390/su16083261

López-Meneses, E., et al. (2025). Educational data mining and predictive modeling in the age of artificial intelligence: An in-depth analysis of research dynamics. *Computers, 14*(2), 68. https://doi.org/10.3390/computers14020068

Martins, R. M., et al. (2024). Machine learning for all!—Introducing machine learning in middle and high school. *International Journal of Artificial Intelligence in Education, 34*(2), 185–223. https://doi.org/10.1007/s40593-022-00325-y

Masci, C., Johnes, G., & Agasisti, T. (2018). Student and school performance across countries: A machine learning approach. *European Journal of Operational Research, 269*(3), 1072–1085. https://doi.org/10.1016/j.ejor.2018.02.031

McCusker, K. A., et al. (2013). Intelligent assessment and content personalisation in adaptive educational systems. In *I2013 12th international conference on information technology based higher education and training (ITHET). 2013 12th international conference on information technology based higher education and training (ITHET)* (pp. 1–7). IEEE. https://doi.org/10.1109/ITHET.2013.6671025

Mehrpouyan, A. (2023). Enhancing online english language and literature classrooms: Effective and practical teaching strategies. *Education and Information Technologies, 28*(4), 4127–4141. https://doi.org/10.1007/s10639-022-11235-w

Na, K. S., & Tasir, Z. (2017). Identifying at-risk students in online learning by analysing learning behaviour: A systematic review. In *2017 IEEE conference on big data and analytics (ICBDA). 2017 IEEE conference on big data and analytics (ICBDA)* (pp. 118–123). IEEE. https://doi.org/10.1109/ICBDAA.2017.8284117

Paleyes, A., Urma, R.-G., & Lawrence, N. D. (2023). Challenges in deploying machine learning: A survey of case studies. *ACM Computing Surveys, 55*(6), 1–29. https://doi.org/10.1145/3533378

Ten Dam, G., & Volman, M. (2004). Critical thinking as a citizenship competence: Teaching strategies. *Learning and Instruction, 14*(4), 359–379. https://doi.org/10.1016/j.learninstruc.2004.01.005

Theissler, A., & Ritzer, P. (2022). EduML: An explorative approach for students and lecturers in machine learning courses. In *2022 IEEE global engineering education conference (EDUCON). 2022 IEEE global engineering education conference (EDUCON)* (pp. 921–928). IEEE. https://doi.org/10.1109/EDUCON52537.2022.9766719

Tsangaratos, P., et al. (2025). Machine learning approaches for real-time mineral classification and educational applications. *Applied Sciences, 15*(4), 1871. https://doi.org/10.3390/app15041871

Vardhan, A., Chauhan, A. S., Saini, S., & Sharma, S. (2026). IoT-driven climate smart agriculture: Adapting to a changing climate for sustainable farming. In S. R. Kumar, N. Gayathri, V. Yendapalli, & A. Rocha (Eds.), *AI-driven IoT solutions in agriculture: New insights. Studies in computational intelligence* (Vol. 1236). Springer. https://doi.org/10.1007/978-3-032-05603-0_5

Vinoth, A., & Datta, S. (2022). Fundamentals of machine learning. In S. Datta & J. P. Davim (Eds.), *Machine learning in industry* (pp. 1–27). Springer International Publishing. (Management and Industrial Engineering). https://doi.org/10.1007/978-3-030-75847-9_1

Wahdan, A., et al. (2022). Early detecting students at risk using machine learning predictive models. In M. Al-Emran et al. (Eds.), *proceedings of international conference on emerging technologies and intelligent systems* (pp. 321–330). Springer international publishing. (lecture notes in networks and systems). https://doi.org/10.1007/978-3-030-85990-9_27

Yekollu, R. K., et al. (2024). AI-driven personalized learning paths: Enhancing education through adaptive systems. In R. Asokan, D. P. Ruiz, & S. Piramuthu (Eds.), *Smart data intelligence* (pp. 507–517). Springer Nature Singapore. (Algorithms for Intelligent Systems). https://doi.org/10.1007/978-981-97-3191-6_38

Yousafzai, B. K., Hayat, M., & Afzal, S. (2020). Application of machine learning and data mining in predicting the performance of intermediate and secondary education level student. *Education and Information Technologies, 25*(6), 4677–4697. https://doi.org/10.1007/s10639-020-10189-1

Challenges and Future Directions: Transforming OBE with Machine Learning

Sumedha Magotra, Hritik Awasthi, Simarpreet Kaur, Vikas Wasson, and Bikram Pal Kaur

Abstract This chapter explores the integration of Machine Learning (ML) into Outcome-Based Education (OBE) to address its implementation challenges and enhance its effectiveness. OBE focuses on clearly defined learning outcomes, promoting student-centric, skill-based learning aligned with industry standards. However, its adoption often compromises the quality of learning experiences due to limited assessments and fragmented feedback systems. To overcome this, the chapter proposes ML techniques that offer personalized feedback, formative assessments, and predictive modeling for student success. Despite their potential, ML models pose challenges like data bias, low interpretability, and privacy concerns. The chapter critically examines how such issues affect learning outcomes and proposes bias mitigation and fairness-aware algorithms to ensure equitable education. Through real-world examples and use cases, it demonstrates how ML enhances adaptive curriculum design, learning analytics, and student performance predictions. The results show improved accessibility, relevance, and inclusivity in education. However, the "black-box" nature of ML remains a barrier for educators to trust and utilize these tools fully. The chapter concludes by highlighting future directions for integrating interpretable and ethical ML solutions into OBE frameworks, encouraging more transparent, inclusive, and outcome-driven learning systems. This interdisciplinary approach redefines education, offering transformative value for students, educators, and institutions alike.

Keywords Outcome-based education · Machine learning · Course outcomes · Program outcomes

S. Magotra · H. Awasthi
AIT-CSE, Chandigarh University, Mohali, Punjab, India

S. Kaur (✉)
UIE-ECE, Chandigarh University, Mohali, Punjab, India

V. Wasson
UIE-CSE, Chandigarh University, Mohali, Punjab, India

B. P. Kaur
The College of New Jersey, Trenton, NJ, USA

D. Vetrithangam et al. (eds.), *Transforming Outcome-Based Education with Machine Learning*, https://doi.org/10.1007/978-981-95-4925-2_9

1 Introduction

Outcome-based education has gained popularity over the last thirty years in response to the growing need for education that produces graduates who are well-prepared for the workforce and possess quantifiable skills. It is based on modern principles of education that are fundamentally different from the traditional model of teaching, wherein the primary teaching process is the transmission of content. OBE emphasizes the importance of learning what is supposed to be acquired, how to utilize the knowledge obtained, how to measure it, and how it is distinctly defined. The OBE system introduces an entirely new method of teaching and learning, whereby the way educational institutions teach and how learning is supposed to occur are vastly different from what has ever been the case before. The measure of education in the OBE system does not hinge upon the time spent in lectures or how much of a text was read; rather, the measure lies in what the students have learned, how that knowledge will be used, and what competencies it builds so that students will be equipped to tackle problems of everyday life (Biggs & Tang, 1999; Felder et al., 2004).

A thought on core modification is that it cannot be theoretical cognition that a student pulls out from an institution, but rather the skills, critical thinking, and problem-solving capabilities necessary to be functional in today's challenging work environments. The demands of industry are transforming rapidly, thanks to technological interventions and the waves of globalization; hence, there is an urgent need for the education systems to respond equally promptly. Against this backdrop, Outcome-Based Education aims to ensure that students acquire the capabilities and attributes highly valued in contemporary work settings (Goldin & Katz, 2010). In parallel, technological advancements, primarily in the realm of Machine Learning (ML), have introduced various robust new techniques to track, assess, and improve student performance. From the perspective of learning analytics that are meant to gather insights on student progress to adaptive curriculum design that shows its capacity to tailor learning experiences to become individually focused, and to build real-time personalized feedback systems that enhance the teaching-learning process, Machine Learning has the potential to transform nearly every aspect of education (Lipton, 2018; Mitchell et al., 2019). However, while ML can augment OBE to facilitate data-driven decision-making, the challenge lies in its integration into educational systems. This chapter discusses the multipronged challenges and opportunities that arise in connecting machine learning with outcome-based education. Key issues, including data quality, algorithm interpretability, fragmented datasets, privacy concerns, and implementation challenges, are examined, along with proposed solutions to address these concerns. In addition, avenues are put under the spotlight wherein the promise is meant to bridge the current gaps towards the formulation of more efficient and fairer educational environments. By engaging in such a discussion, the parties involved—educators, researchers, and policymakers —can be engaged, as it not only points out technical challenges but also considers pedagogical and ethical questions necessary for enabling a well-rounded transformation of education.

The structure of this chapter is as follows. In Sect. 2, the evolution and fundamental principles of Outcome-Based Education, as well as how present Machine Learning methods are being employed in the educational environment, are put into perspective. Section 3 showcases and discusses real-life case studies that analyze CO-PO attainment, achievement levels, and the actions taken to improve them, all of which align with the university's vision and mission statements. Section 4 presents the ongoing hurdles of DLL integration into OBE, including subsections that discuss data structure quality, algorithm transparency, data partitioning, and privacy concerns. Then, in Sect. 5, some remedial actions to address these hurdles are presented, including recommendations on data governance, the role of Explainable AI (XAI), and change management within the institution. Section 6 presents opportunities that include dynamic curriculum development, adaptive systems for evaluation and feedback, developing research initiatives, and informing policy directions. Section 7 presents cases from around the world as best practices, showcasing the promise and also the challenges that ML-infused outcome-based systems bring. Lastly, Section 8 presents the conclusions and outlines a research agenda for the next phase of development. Our approach is interdisciplinary throughout the chapter. By considering the technical, educational, ethical, and organizational perspectives, a roadmap for the future of ever-changing education is provided. The aim is to provide great encouragement to educators and administrators to identify the best opportunities for Machine Learning, while also preventing the worst. Only in such a balanced and considered approach can one envisage a reality where the pillars of the Outcome-Based Education platform, based on twenty-first-century technology, are realized to their fullest potential.

2 Background Study

The assumption underlying OBE stems from a critique of traditional education for delivering content but not proving that competency has occurred. Traditional education stresses memorization, while ensuring that students have acquired meaningful knowledge and skills doesn't seem to be its top priority (Biggs & Tang, 1999). OBE, on the other hand, focuses on student outcomes, ensuring that the knowledge a student gains during their education is both academically sound and applicable to the professions that utilize this knowledge. Such a shift in paradigms led to the adoption of active learning approaches and the alignment of educational aims with occupational standards.

Felder and Brent claim that, since the second half of the twentieth century, OBE has significantly influenced technical education, particularly in fields such as engineering and medicine, where learning objectives must adapt to the ever-changing demands of the workplace. Throughout history, some changes in the development of OBE have been very significant. Active Incorporating Learning: Active Learning is the practice of actively incorporating learning, as opposed to setting learning environments where students start in passive roles and are then directed by lecturing. Under this style, methodologies such as project work and problem-solving exercises serve as examples that make a practical contribution, enabling students to gain hands-on experience. The formative assessment is valued more in this systems model of OBE than are high-stakes examinations because of their provision for continuous feedback, allowing for modifications to be made to the curriculum. On-the-industry alignment: The unification of OBE to ensure graduates are meeting current job demands, factoring in the industry voice (Goldin & Katz, 2010). These include personal learning pathways: Due to the various types of learning needs, OBE intends to introduce personalized learning pathways by applying adaptive learning approaches. While these advances have bolstered the effectiveness of OBE, challenges still exist in implementing it robustly, especially in terms of the timely scalability of these principles across diverse educational contexts. The intersection of machine learning (ML) with OBE presents both challenges related to bias, interpretability, and data ethics, while offering new opportunities for fine-tuning and optimization.

The advent of machine learning in academic settings has transformed the way students' learning is assessed and optimized. Machine learning algorithms extract information that enables one to improve the learning experiences of students by analyzing volumes of data that contain test scores, attendance patterns, and engagement metrics (Siemens, 2013). The following subsections discuss the role of machine learning in learning analytics, adaptive curriculum design, formative assessment, predictive analytics, and administrative optimization. Learning analytics can be defined as a data-driven approach to understanding and enhancing the learning process. The work of Papamitsiou & Economides, 2014) characterizes machine-learning-driven analytics in learning as the identification of trends within student performance, allowing for a form of proactive intervention. Using clickstream data from online learning platforms, ML models can reveal the topics that students struggle with the most and refine the instructional strategies accordingly (Romero & Ventura, 2020).

Commercial curricula are essentially static, susceptible to the differential learning paces of their charge. Singh et al. (2024) looked into the contribution of ML to developing adaptive curricula in which instructional content fluctuates based on individual student performance. Such personalization—a sure engagement booster—also ensures mastery of learning objectives before students advance to the following concepts. The other significant development of educational data mining is the use of ML in formative assessments. Intelligent tutoring systems leverage ML algorithms to provide real-time feedback to students regarding commonly held misconceptions and tailored remediation (Baker & Inventado, 2014). While traditional summative assessments are typically administered at the end of a learning cycle to provide feedback, ML-based formative assessments enable continuous learning through the real-time adjustment of instructional strategies. Thus, ML in predictive analytics can enable forecasting on student performance and retention risks, allowing for early intervention strategies by instructors. Research by Ferguson (2012) demonstrated that predictive models derived from historical K-12 academic data could effectively identify at-risk students. Predictive analytics informs early intervention programs, which, in turn, have been shown to lead to improvements in rates of student success and overall learning outcomes (D'Mello et al., 2019; Tsai, 2024).

Besides the direct improvement in learning, ML plays a role in easing different administrative activities in educational institutions. O'Neil (2016) noted that scheduling, resource management, and even student services optimization can be achieved through efficiency within the institution by utilizing ML-driven methodologies (O'Neil, 2016). However, using ML inside administrative systems raises ethical concerns, such as data privacy and bias, that deserve significant consideration. Several challenges in machine learning for outcome-based education have emerged, which need to be addressed for machine learning to be transformational; thus, ML model bias is one of the most critical issues to be discussed in education systems. As pointed out by Fleuren et al. (2018), biased training datasets inevitably lead to unfair learning recommendations, placing specific student demographics at a greater disadvantage. Various bias mitigation strategies have been proposed, including fairness-aware algorithms, to address this predicament (Mitchell et al., 2019). Model interpretability remains a high-priority challenge in the context of OBE. In his work, Lipton (2018) argues that the opacity associated with ML methods makes it difficult for educators to understand how predictions were generated in a specific case. Limited interpretability blocks trust and adoption from wave educators; therefore, explanations in terms of explainable AI models must be developed. Plausibly, ethical problems related to student data, particularly concerning privacy and data security, have come sharply into the limelight. According to Baeck (2015), standardized protocols considering data collection and storage are necessary to maintain the confidentiality of students. Educational institutions should exercise due diligence in implementing adequate data governance to mitigate potential data breach risks and unauthorized access.

The ML-focused OBE could become a reality once the existing challenges are figured out with the help of new advancements in AI and data analytics. Research depends on various aspects that may shape the trends to guide the future. Model interpretability efforts are considered of the most significant importance to include teacher trust and uptake. Recent research on explanation frameworks of XAI that incorporate methods to make decision processes explicit has gained momentum (Lipton, 2018). Research on intelligent algorithms that include fairness and equality considerations is in progress: such endeavours would rather aim to reduce bias from learning models driven by ML. Mitchell et al. (2019) proposed Model Cards for reporting the fairness of AI, which provide standard definitions for brevity and clarity—advances in AI, as seen in Google and Andrew G. Hutch's SWOT Analysis. Powered personalization will allow even greater responsiveness in curricula for OBE. Research on educational systems based on reinforcement learning suggests that dynamic cha…In terms of learning pathways, they can be customized for the individual needs of a learner (Baker & Inventado, 2014).

While machine learning will drastically change Outcome-Based Education through its advanced capabilities in data-driven decision-making, personalized learning, and predictive analytics, it also introduces inherent challenges concerning databases, model interpretability, and the ethical implications of its use that need to be addressed before more equitable and effective implementations can be undertaken. The focus should shift to research in the area of improving the explainability of AI tools, developing effective bias remediation models, and further implementing AI-driven personalization to refine OBE paradigms. By addressing these challenges, ML will make the next step in transforming education into a truly student-centered and outcomes-driven system.

3 Case Studies

The format of outcome-based education (OBE) facilitates alignment of student learning outcomes with institutional objectives. This section will share case studies of technical courses to highlight the implementation of OBE, while demonstrating student achievement of Course Outcomes (COs) and Program Outcomes (POs). These case studies illustrate the interplay between the institutional vision, institutional mission, COs, and POs, and also emphasize the critical role that assessment tools and strategies for continuous improvement can take.

3.1 OBE Framework

As shown in (Fig. 1), the foundation of OBE lies in the clear articulation of an organization's vision and mission, which inform instructional methods, curriculum design, and student experiences.

For example, the vision may focus on developing skilled professionals capable of addressing societal and industrial challenges while promoting sustainability. The mission, in turn, could emphasize delivering quality education in engineering and related fields, fostering innovation, inquiry, and ethical practices in academic endeavors, and equipping students with technical expertise, business acumen, and a commitment to lifelong learning.

Building on these foundational elements, Program Educational Objectives (PEOs) and Program Outcomes (POs) are defined to provide more specific educational goals. POs describe the skills, dispositions, and knowledge that students should have by the time they graduate, while PEOs are broad objectives that reflect graduates' long-term professional and career goals. Degree outcomes further define expectations and offer measurable indicators of student proficiency in particular fields. The cognitive domains of applying, analyzing, evaluating, and creating—part of Bloom's Taxonomy—are utilized to ensure structured evaluation and facilitate mapping to align PEOs with POs appropriately.

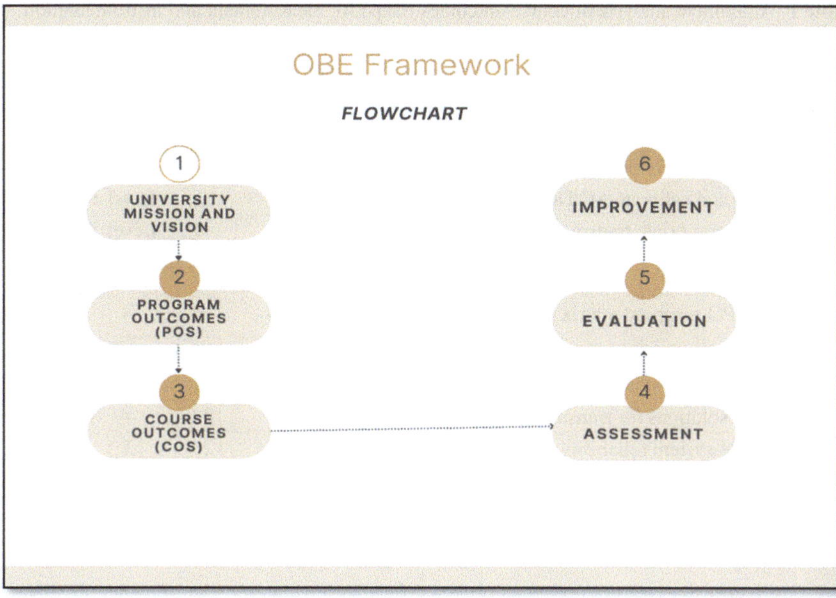

Fig. 1 OBE framework

3.1.1 Case Study 1: Assessment of CO and PO Achievement in Fundamental Engineering Courses

The aim of implementing had always been to finalize the levels of attainment of both Program and Course Outcomes for an engineering core subject. The implementation included both indirect and direct methods of evaluation. The measures for the study included designing 6 COs that targeted various analytical abilities, such as 'Estimating Error', 'Electrical Instrumentation', and 'Calibration of systems and instruments'. The COs got mapped to chosen POs that addressed Problem Solving (PO2), Engineering Knowledge (PO1), and Ethical Behaviour (PO8). While the final semester examinations were evaluated using HOT (High-order Thinking) Criteria, each separate course was assigned to assess the internal exams, including mid-semester exams, quizzes, and assignments. Multiple feedback surveys and course exit questionnaires were used to gauge students' perceptions of their learning experience and educational attainment. Furthermore, inputs from Alumni were also taken to substantiate the industry-based application of the POs. The results showed that CO1 and CO3 were able to achieve high attainment levels; however, CO2 only depicted average attainment, suggesting that more focus is necessary in that area. The CO statements and their respective Attainment Levels are mentioned in Table 1.

Table 1 Attainment analysis of course outcomes (COs) for measurement and instrumentation

CO Statement	Attainment Level (%)	Achievement Level
CO1: Explain the working principle of measuring instruments.	75%	Level 3
CO2: Analyze measurement system errors.	68%	Level 2
CO3: Design and calibrate AC/DC bridges.	80%	Level 3
CO4: Select appropriate transducers for measurement.	72%	Level 3

Table 2 CO-PO mapping matrix for power system analysis

CO Statement	PO1	PO2	PO3	PO4	PO5
CO1: Analyze power flow models.	3	3	–	2	–
CO2: Design solutions for transmission networks.	3	2	3	–	2
CO3: Evaluate system reliability.	3	–	2	3	–

3.1.2 Case Study 2: Comprehensive PO Attainment in Advanced Engineering Courses

This case study aimed to assess Program Outcome (PO) attainment for Power System Analysis, an advanced subject, through both direct and indirect evaluation methods. Building technical skills like power flow modeling and system reliability evaluation was the goal of the course. To ensure that each Course Outcome (CO) aligned with institutional objectives, it was carefully mapped to pertinent Program Outcomes, including Engineering Knowledge (PO1), Problem Analysis (PO2), and Communication Skills (PO10). The CO-PO mapping matrix for this subject is shown in Table 2, which demonstrates the strength of alignment between specific course outcomes and targeted POs.

The level of PO attainment and corresponding improvement actions for the academic session 2023–2024 are presented in Table 3. Based on the evaluation, workshops, case studies, projects, and simulation tools were introduced to improve learning outcomes. Ethical decision-making was also integrated through dedicated modules.

The computation of overall CO attainment was based on a weighted average where 80% weight was given to direct assessment methods and 20% to indirect assessment, following standard attainment calculation procedures. Direct assessment included internal exams and projects, while indirect assessment consisted of course exit surveys and feedback from students. The overall CO attainment for this subject is shown in Table 4. Please note that the values in this table are sample values used for demonstration purposes and should be validated using actual assessment data during implementation.

The case study provided valuable insights into the process of aligning course outcomes with institutional objectives. The structured CO-PO mapping ensured that learning goals were clearly defined and measurable. Direct methods, such as exams

Table 3 PO attainment levels and improvement actions (2023–2024)

PO	Attainment Level	Actions for Improvement (2023–24)
PO1	3	Organize workshops on advanced engineering fundamentals.
PO2	2	Conduct case studies on real-world system failures.
PO3	2	Introduce collaborative projects and practical assignments.
PO4	3	Include simulation tools for power system problem-solving.
PO8	2	Create modules emphasizing ethical decision-making.

Table 4 Overall CO attainment computations

CO Statement	Direct Attainment (%)	Indirect Attainment (%)	Overall CO Attainment (%)
CO1: Analyze power flow models.	80%	20%	76%
CO2: Design solutions for networks.	70%	20%	68%
CO3: Evaluate system reliability.	75%	20%	72%

and projects, offered a quantitative evaluation of technical abilities, while indirect feedback from students and alumni added qualitative depth to the review. Areas showing lower attainment led to specific corrective actions such as curriculum revisions and the introduction of advanced tools for hands-on learning. Additionally, ethical dimensions were gradually incorporated into the technical content to prepare students for their professional responsibilities better.

4 Challenges in Integrating Machine Learning with Outcome-Based Education

The combination of theory-based Machine Learning and Outcome-Based Education (OBE) is seen as promising, but it is also considered challenging. Many-layered problems need to be solved before the benefits of advanced technologies can be fully used. In this section, the main challenges faced by educators and administrators are discussed. These include technical problems, ethical concerns, and organizational issues that must be carefully handled.

4.1 Data Quality: The Foundation of Trustworthy Analytics

The effectiveness of any ML system relies heavily on the type of data it receives. This means that high-quality data is critical, or, as many experts would say, "garbage in, garbage out." There remain many factors that compound the problem of data quality, among them are:

4.1.1 Inherent Bias in Datasets

Educational data can be affected by bias at multiple levels. Systemic issues, such as socioeconomic disparities, historical underrepresentation, and varying access to quality education, can skew the data. For instance, the data is predominantly extracted from urban schools; thus, ML models may not perform well when transferred to rural schools. This bias may lead to poor decisions that could further widen the existing inequities present in the education system. Thus, active sampling is essential, including the deployment of active bias correction strategies and the collection of diverse data sources.

4.1.2 Fragmentation of Data Sources

Educational data is often scattered across various systems, such as online learning platforms, examination databases, student records, and even extracurricular activity logs. Bringing all this data together into a single, unified dataset is a complex task. Differences in data formats, collection frequencies, and measurement standards can cause inconsistencies in machine learning models. One possible solution is to use centralized data warehousing and adopt universal data standards (Veale & Binns, 2017).

4.1.3 Inconsistencies in Data Collection and Reporting

Different institutions follow different methods to collect and report student data. Even within the same institution, inconsistencies may occur due to differences in grading practices among teachers and the use of various assessment formats. Since consistent and dependable data are essential for efficient learning and prediction, these discrepancies may impact the precision of machine learning models. Strong data cleaning methods, data normalization, and appropriate error-checking procedures are necessary to address this and ensure the accuracy and consistency of input data (Celik et al., 2022).

4.1.4 The Role of Big Data

As educational environments become increasingly digitally connected, particularly with the proliferation of online learning platforms, the amount of available data has increased dramatically. Though it has the potential to yield profound insights, this "big data" also presents storage, processing, and analytics challenges. It requires sophisticated data mining techniques, scalable infrastructure, and the ability to distinguish meaningful signals from noise to effectively utilize big data.

4.2 Interpretability and Transparency of ML Models

One of the most well-known problems with machine learning is the opaqueness of many complex algorithms. Understanding how a model arrived at a particular prediction becomes crucial when educators depend on these models to make critical decisions.

4.2.1 The "Black Box" Problem

Many advanced machine learning methods, such as deep neural networks, are commonly referred to as "black boxes" because users are unable to see how they make their decisions. The lack of transparency in these systems makes it difficult for teachers to trust them, especially when important decisions are involved. Teachers might be hesitant to step in until they have a firm grasp of the factors that led to the prediction, for example, if an ML model predicts that a student might perform poorly academically without offering a justification.

4.2.2 Challenges in Explainable AI (XAI)

XAI has emerged as a trustworthy approach to increase the transparency and trustworthiness of ML systems. However, this field is still in an evolving phase when it comes to education. There is still room for improvement, and research is being conducted on tools such as model cards, Relevance Maps, and Sensitivity analysis in a real classroom environment. These AI techniques can be somewhat complex for newcomers; educators need to understand how to effectively utilize this information to teach students and enhance their learning skills. These problems can pose a barrier to both teaching and learning.

4.2.3 Fostering Credibility through Insight

Performing operations in ML is comparatively more optimized, which has significantly improved trust. If given proper guidance and a clear explanation, educators and administrators are more inclined to embrace these technologies. It helps the educators by providing user-friendly dashboards and visual representations, ensuring that the interpretability of the data is prioritized. Establishing a transparent, trustworthy, and understandable workflow fosters greater confidence and emphasizes the effective use of AI in education (Sanusi et al., 2024).

4.3 Fragmentation *and Integration of Educational Data*

Fragmentation of data is not merely a technical concern; it significantly affects the overall performance of the ML system, including accuracy and fairness, which can alter the results.

4.3.1 Siloed Educational Data

In many institutes, data is segregated across different databases or stored in other systems for individual programs. The unorganized separation of data may lead to incomplete analytics and integration difficulties. This disjoint structure alters the ML performance; ML models fail to detect critical key correlations or miscalculate the outcome due to the lack of unified data. Discussing and addressing this problem necessitates the adoption of standardized data formats and the integration of multiple datasets into a single repository (Wang, 2024).

4.3.2 Advanced Data Integration Techniques

To ensure the reliable and seamless data integration across all the institutes, standardized APIs and advanced metadata management techniques play a crucial role. It tracks the data generation and preprocessing around the institutes. An ML-driven tool helps in several ways, including automated data cleaning and minimizing inconsistencies, which ensures more accurate and dependable analytics (Levin and Tomsk State University of Control Systems and Radio Electronics, 2024; Kumari et al., 2023).

4.4 Data Privacy, Security, and Ethical Considerations

The ML in education comes with certain restrictions and responsibilities; data privacy and ethical workflow are essential considerations.

4.4.1 The Imperative of Privacy Protection

Institutional data is often very personal and sensitive, as it contains the personal details of both students and staff. Policies with data protection regulations, such as the GDPR and FERPA, are crucial to implement for data safety. Securing data alone is not sufficient; institutions must also implement data processing methods that uphold confidentiality and anonymity. It involves enforcing strict access control and applying encryption protocols to protect the privacy of all individuals involved.

4.4.2 Security in the Digital Age

The high dependency of educational institutions on cloud computing and interconnected systems makes the system vulnerable and creates some loopholes. The risk of cybersecurity breaches increases. These breaches not only expose sensitive personal data but also shatter the trust of people in the ML systems that support educational processes. Therefore, implementing strong cybersecurity practices is crucial. It involves regular checking, performing penetration tests, and providing ongoing training to the onboard employees. Likewise, the implementation of blockchain technologies for data integrity is being explored as a way to ensure that data remains tamper-proof and secure in long-term repositories.

4.4.3 Ethical Dilemmas in ML-Driven Education

Beyond privacy and security, ethical considerations are also significant. How should algorithms balance the pursuit of predictive accuracy against the need for fairness? How can we ensure that data-driven decisions do not reinforce historical inequalities? These are not merely technical challenges—they are ethical imperatives that require input from educators, ethicists, policymakers, and the broader community. Establishing ethics review boards and developing comprehensive guidelines for the responsible use of ML can help navigate these dilemmas (Cai et al., 2024).

4.5 *Organizational and Implementation Challenges*

Implementing ML-enhanced outcome-based education systems often encounters resistance at the organizational level.

4.5.1 Cultural Resistance in Educational Institutions

Many educational institutions are traditional in nature, which makes it difficult to imagine implementing radical technological changes within such environments. Teachers and administrators who have long depended on conventional methods of assessment may view machine learning not as a helpful advancement but as an unfamiliar and potentially disruptive tool. Therefore, what is needed is not only a shift in technology but also a transformation in mindset—a broader cultural shift that embraces evidence-based decision-making and continuous improvement (López-Meneses et al., 2025).

4.5.2 Skill Gaps and Professional Development

Creating and sustaining complex ML systems requires an inordinate amount of technical expertise. More often than not, teachers are not trained in data analytics or ML techniques, and technical support may be limited within schools. A comprehensive professional development program, directed toward teachers, administrators, and IT staff alike, can help fill this gap. Training programs, workshops, and collaborative research initiatives must be a significant part of the institutions' investments to build skills from the ground up.

4.5.3 Financial and Logistical Constraints

Adopting ML technologies is resource-intensive. Purchasing hardware and software infrastructure, training programs, and hiring specialized personnel can pose significant financial barriers. Other logistical concerns, such as integrating new systems with legacy technologies, may also delay implementation. Hence, institutions must begin to explore innovative models of financing, public-private partnerships, and phased rollout options to overcome such challenges (Marks & Al-Ali, 2022).

5 Strategies for Overcoming Current Challenges

Institutions should adopt a multi-pronged approach to realize the potential of ML-enhanced OBE fully. We outline here strategies that address the above technical, operational, and ethical challenges.

5.1 Enhancing Data Quality and Mitigating Bias

Improving data quality is pivotal for the effective deployment of any ML system. The following strategies offer a roadmap for a data-centric approach:

5.1.1 Standardization of Data

Institutions should collaboratively establish universal data collection standards. These standards encompass a wide range of topics, including data formats, storage protocols, assessment methodologies, and grading systems. Uniform data standards reduce the likelihood of inconsistencies and enable effective integration. For example, consistent attendance metrics, standardized digital transcripts, and uniform score scales can all be used to build a robust data system.

5.1.2 Regular Data Audits

Institutions must regularly audit datasets to detect and address biases or errors quickly. Data quality should be assessed periodically to ensure accuracy, completeness, and compliance with predetermined standards. ML techniques can be applied in audits to improperly draw attention to the problem and hasten the adoption of corrective actions.

5.1.3 Incorporation of Fairness-Aware Algorithms

To reduce bias and ensure equitable results, fairness-aware machine learning algorithms must be used in conjunction with strategies such as adversarial debiasing, resampling, and fairness constraints—as well as ongoing monitoring across student groups.

5.1.4 Inclusive and Diverse Data Sourcing

In addition to making technical changes, educational institutions should create proactive plans to gather a range of data. They should collaborate with many schools, districts, and countries to ensure that their datasets reflect a diverse range of student backgrounds. Such diversity will not only significantly improve model accuracy but also advance social equity.

5.2 Advancing Explainable AI (XAI) for Enhanced Trust

Strong steps must be taken to make the "black box" more transparent, thereby building confidence in ML systems.

1) ***Development of User-Friendly Model Cards***
Short documents, known as "model cards," outline the problem the machine learning model is trying to solve, how its performance is assessed, its limitations, and for whom the model is intended. Having model cards for each algorithm that has been deployed, institutions can provide teachers with documentation that explains what such a model is and how it works, enabling trust and acceptance (Hassija et al., 2023).

2) ***Interactive Dashboards and Visualizations***
Visualization tools that transform raw data and ML outputs into clear, actionable insights are crucial. Dashboards, interactive reporting tools, and data visualization software can help educators quickly understand key performance indicators and model decisions. These interfaces must be designed in consultation with end users—teachers and administrators—to ensure that they meet the practical needs of the educational setting.

3) ***Collaborative Development Initiatives***
Special workshops, pilot projects, and joint ventures between machine learning experts and teachers may contribute to the establishment of a culture of co-development. Engaging teachers from the initial stages of system design can help ensure that the outputs, explanations, and guidelines eventually produced are significant within the real classroom environment. Such interdisciplinary cooperation is crucial for bridging gaps in understanding and achieving more effective outcomes.

5.3 Integrating and Harmonizing Fragmented Educational Data

Overcoming data fragmentation requires both technical systems and organizational policy changes

1) ***Establishing Centralized Data Repositories***
Educational institutions can significantly benefit from creating centralized data warehouses. Repositories, such as data warehouses and data lakes, store data from multiple sources in one place for more effective analysis. Cloud solutions that have integration protocols and access controls are capable of doing this effectively.

2) ***Implementing Advanced ETL (Extract, Transform, Load) Processes***
ETL procedures are essential for integrating data. These consist of gathering data from multiple sources, converting it into a standard format, and putting it into a shared warehouse. Automating anomaly detection and enhancing data cleaning for large, diverse datasets necessitates sophisticated ETL tools that utilize machine learning algorithms.

3) ***Metadata and Data Provenance***
Through transparent data provenance, an audit of potential bias or errors becomes possible. Transparency and reliability are reinforced by maintaining thorough metadata on data collection and processing.

5.4 Ensuring Data Privacy, Security, and Ethical Use

It is genuinely required to use data in an ethical manner, which necessitates a diversified approach.

1) ***Robust Anonymization and Encryption Protocols***
It is necessary to incorporate state-of-the-art methodologies to help maintain the anonymity of student data. Methods such as tokenization, encryption, and pseudonymization must be used to maintain the confidentiality of personal data. Comprehensive control access and regular updates of these security protocols help protect sensitive data from any unauthorized access.

2) ***Transparent Data Use Policies***
Institutions must design and publish clear, easy-to-understand policies that describe how data will be accumulated, stored, and shared. Transparent data policies foster trust among stakeholders, including parents, teachers, and students, as they clearly outline the benefits and uses of data.

3) ***Ethics Review Boards and Continuous Monitoring***
Ethical review panels should be established to keep a check on ML initiatives in education. The panel will have the authority to make desired adjustments if any element does not adhere to any ethical policies.

5.5 Change Management and Building a Technology-Driven Culture

Changes in organizations are necessary to integrate ML accurately. They can be as follows:

1) ***Comprehensive Professional Development Programs***
The success of technology-driven educational advancements is based on the involvement of all stakeholders. To close knowledge gaps, institutions should provide funding for professional courses that focus on digital literacy and ethics. They can also utilize conferences and certifications as an additional method.

2) ***Stakeholder Inclusion and Communication***
Early faculty, staff, and student involvement is essential for effective change management, and clear communication about goals, advantages, difficulties, and deadlines is also necessary to minimize resistance and promote teamwork.

6 Future Directions in ML-Enhanced Outcome-Based Education

This section examines the future directions for transforming education. Deeper ML-OBE integration will be driven by changing technological and educational needs.

6.1 Dynamic and Evolving Curriculum Design

Education's future will be shaped by its ability to adapt the curriculum in real-time to meet industry and student demands.

1) ***Real-Time Update of Learning Outcomes***
Academic courses must remain in sync with industry demands and societal trends. A system can use real-time student performance data and enhanced ML to recognize trends and also suggest updates in the curriculum.

2) ***Personalized Learning Trajectories***
Similarly, future systems will use ML and big data to personalize learning paths by making changes in the content and pace, which will allow teachers to adjust based on each student's requirements and strengths.

6.2 Next-Generation Explainable AI and Decision Support

The gap between sophisticated ML models and their real-world implementation will close as XAI advances.

6.2.1 Interactive Decision-Support Dashboards

Interactive, real-time dashboards will be a key feature of future machine learning systems, enabling teachers to investigate the variables that affect model results. Non-technical users will find it easier to comprehend and implement the recommendations thanks to these visual tools, which present data in understandable formats, such as flowcharts, heat maps, and graphs. A dashboard, for example, could illustrate how differences in extracurricular activities, assignment grades, and attendance all affect the results of a predictive model regarding student success.

6.2.2 Enhanced Model Transparency via Multidisciplinary Methods

It is anticipated that new techniques for model interpretability will result from the cooperation of ML specialists, educators, and cognitive scientists. Subsequent studies could lead to algorithms that relate statistical data to pedagogical theories and convey decision logic in a narrative form. Such diverse endeavours may yield "explanation modules" that simplify intricate model reasoning into practical classroom knowledge.

6.2.3 Open-Source Collaborative Frameworks

The establishment of open-source platforms for learning machine learning models will promote increased openness and the exchange of best practices, ensuring that new solutions are thoroughly tested and broadly applicable. These collaborative frameworks will enable institutions worldwide to contribute to and enhance evidence-based machine learning solutions.

6.3 Adaptive Assessment, Real-Time Feedback, and Virtual Tutoring

Machine learning is revolutionizing assessment by enabling continuous, adaptive feedback.

6.3.1 Revolutionizing Formative Assessment

Traditional summative examinations are being replaced by ongoing formative assessments powered by machine learning. In the future, assessments will be seamlessly integrated into the learning process as interactive exercises, simulations, or online quizzes that adjust based on student responses. When a learner is having trouble understanding a concept, these systems can identify it and provide more practice opportunities and tailored feedback immediately.

6.3.2 Virtual Tutors and Intelligent Educational Assistants

ML-powered virtual instructors are expected to become a standard feature of schooling. These intelligent assistants will actively guide students through personalized learning routes, in addition to providing answers. By monitoring progress in real-time and adapting their teaching strategies accordingly, virtual tutors can provide one-on-one help that closely mimics the quality of top-notch human tuition. This enables the reach of large audiences with customized instruction (Tan et al., 2022).

6.3.3 Embedding Adaptive Systems in Learning Management Systems

In the future, ML technologies will be seamlessly integrated into all-inclusive Learning Management Systems (LMS). Teachers will be able to monitor students' development in several areas, including cognitive, behavioural, and affective domains, thanks to this integration. A continuous feedback loop that improves instruction and learning outcomes will be created by using the assessment data to make real-time adjustments to learning objectives and activities.

6.4 Building Collaborative and Equitable Educational Ecosystems

Education is fundamentally a human enterprise, and the full potential of ML-driven systems will be realized only in equitable and collaborative ecosystems.

6.4.1 Multi-Stakeholder Partnerships for Systemic Change

Isolated efforts are unlikely to result in a reform of education. Long-term success, however, relies on collaborations among educators, legislators, business executives, developers, and community members. These stakeholders can establish policies, secure financing, and set common standards that support technology-driven teaching while maintaining accountability and equity by working together in collaboration.

6.4.2 Focused Equity Interventions and Fairness-Aware Policies

Future machine learning systems must be designed with equity as their top priority. Institutions should unavoidably collect reviews of machine learning systems to ensure that all students receive the assistance they deserve. Achievement gaps can also be reduced by maintaining proper records of any discrimination and nonadherence to Fairness-Aware Policies.

6.4.3 Research Centers and Policy Think Tanks

To test models, share innovations, and work with policy think tanks to translate research into scalable, actionable policies, multidisciplinary research centers focusing on ML incorporated education must be created.

7 Case Studies and Ideal Solutions

All the case studies and real-life instances are of utmost importance to showcase the benefits and obstacles of integrating ML with OBE.

7.1 Developing Platforms for Adaptive Learning

Leveraging the use of simulation data, a well-known engineering college implemented an ML-based learning platform that greatly enhanced the performance of its students and their satisfaction. By maintaining the improvement criteria, support from faculty increased. This further highlighted the importance of continuous monitoring, improvement, and collaboration.

7.1.1 AI-Powered Formative Evaluations in Medical Education

To enhance the clinical reasoning of students, a renowned medical college integrated an AI-based assessment. These were used to simulate patient interaction, providing instant feedback. Despite initial obstacles, the program could become a standard model with the help of the faculty. This ultimately led to a need for substantial training in upcoming technologies.

7.1.2 Equity-Enhanced Programs in Urban School Districts

An ML system was implemented in a large urban school to analyze data from multiple resources, including test results, attendance records, and student behavior observations, to identify students who may be academically at risk. After the system identified students needing more attention, the school implemented focused interventions, namely, providing additional learning modules, tutorial and mentor programs, and customised tutoring, which helped reduce the achievement difference between demographic groups over several academic years. This was only possible with the combined efforts of parents, educators, and the school system in establishing a learning environment that was supportive and conducive to student success.

7.2 *Solutions and Guidelines for Implementation*

After analyzing various institutions, the best practices found among them are as follows:

7.2.1 Phased Implementation

Firstly, through pilot programs, institutions can gradually incorporate machine learning, allowing for more space for future advancements (Suthar et al., 2022). These early stages facilitate the early identification of problems and the development of appropriate solutions. Obtaining timely input from teachers and students is crucial to refining the system before it is implemented on a larger scale.

7.2.2 Investment in Professional Development

Regular workshops and training sessions in digital literacy, data analytics, and machine learning principles equip staff members and educators to engage with new technologies confidently. A more integrated and prosperous implementation process is supported by the creation of diverse teams comprising educators, data scientists, and IT specialists (Nafea, 2018).

7.2.3 Collaborative Frameworks

Institutions can share best practices and create standardized procedures for data integration and analysis by forming partnerships that include key stakeholders, such as parents and community members, thereby ensuring openness and inclusivity.

7.2.4 Ethical Practices

The implementation of specialized ethics committees fosters collaborative innovation by enriching community knowledge bases, sharing effective solutions, and supporting cooperative research projects to address common challenges effectively. It also enables ongoing evaluation and maintains equity and responsibility in AI/ML applications by utilizing open-source tools (Jagwani, 2019).

7.3 Implications for Educators

The role of educators is directly impacted by the incorporation of machine learning into outcome-based education, leading to essential consequences that include:

1) ***Shifting Educational Roles***
Increasingly, educators will assume the role of learning facilitators rather than content providers. As ML systems increasingly handle routine assessments and fundamental data analysis, teachers have more time to engage in mentoring, critical thinking conversations, and innovative pedagogy. This aligns with the overarching objectives of OBE, which aim to establish dynamic, student-centered learning environments.

2) ***Enhanced Decision-Making***
Teachers will be able to access real-time, detailed data on student performance. Teachers can create customized interventions and track their effects over time by fusing their professional judgment with practical ML-driven insights. Teachers are empowered to make evidence-based decisions that enhance learning outcomes thanks to this data-centric approach (Y et al., 2024).

3) ***Navigating Ethical and Privacy Concerns***
Educators must be aware of the ethical concerns and outcomes that ML integration in education might lead to, to ensure that the technology is used responsibly. It is also essential to ensure that all technological advancements in the field of education comply with data privacy laws.

4) ***Continuous Collaboration and Lifelong Learning***
With ongoing research and development in ML, professionals involved in education must be devoted to continuous learning with respect to fast, ever-evolving technological changes.

8 Conclusions and Future Scope

To summarize, combining OBE and ML is a revolutionary development. Institutions can meet the needs of their students by implementing more balanced evaluation systems, providing timely and consistent support, and creating personalized learning paths through the use of data-driven approaches. This chapter has extensively explored the potential benefits and challenges of integrating ML with OBE. Inconsistent data practices and difficulties with model interpretability are two major technical obstacles that have been identified. Adoption-related organizational, cultural, and ethical factors have also received attention simultaneously. The significance of comprehensive approaches that incorporate technological innovation, ongoing capacity building, and robust policy frameworks has been emphasized as a means of overcoming these obstacles. Multi-stakeholder implementation has been promoted as a way to guarantee responsible and inclusive outcomes.

A thorough understanding of how to successfully incorporate machine learning into OBE systems has been developed. There have been suggestions for improving assessment techniques and personalizing educational experiences. Along with the significance of creating explainable systems, issues such as algorithmic bias, fragmented datasets, and opaque decision-making have also been addressed. Examining both direct and indirect methods for evaluating Program and Course Outcomes in accordance with learner progress indicators and more general institutional goals has also been covered. These conversations have helped lay the groundwork for upcoming studies and policy initiatives aimed at developing student-centered, data-driven teaching methods.

Despite the notable advancements, several issues still require attention; further research is needed on data quality in educational settings, particularly regarding the causes and consequences of bias. Newer methods must be explored in the field of XAI to assist non-professional users in understanding the complex ML models. To understand the effects of tools like model cards and interpretability dashboards, real-life case studies should be used. Studies should analyze the satisfaction ratings and long-term outcomes of students following customized learning paths. Curriculum designs must increasingly include real-time feedback methods. To address data privacy and ethical concerns, interdisciplinary models that safeguard these are necessary, while also formalizing best practices for the fair and secure use of student data in ML settings. A notable difference in modern-day education is the accumulation of adaptive technologies and outcome-based learning, which lures inclusive, moral research as well as continuous innovation to create ever-changing systems. Enhanced training in ML, data analytics, and digital fluency is also necessary for educators, and the promotion of collaboration models within organizations.

Careful judgment is needed at the intersection of machine learning and emerging technologies, such as IoT, blockchain, and AR/VR, with projects that showcase their potential to revolutionize education.

References

Baeck, P. (2015). Data harmonization and standardization in the digital age: Educational implications. *Journal of Information Systems, 29*(1), 45–62.

Baker, R. S., & Inventado, P. S. (2014). *Educational data mining and learning analytics* (pp. 61–75). Springer eBooks. https://doi.org/10.1007/978-1-4614-3305-7_4

Biggs, J. B., & Tang, C. S. -K. (1999). *Teaching for quality learning at university: What the student does* [online]. Available from http://ci.nii.ac.jp/ncid/BA86633208

Cai, C., Zhu, G., & Ma, M. (2024). A systemic review of AI for interdisciplinary learning: Application contexts, roles, and influences. *Education and Information Technologies*. https://doi.org/10.1007/s10639-024-13193-x

Celik, I., Gedrimiene, E., Silvola, A., & Muukkonen, H. (2022). Response of learning analytics to the online education challenges during pandemic: Opportunities and key examples in higher education. *Policy Futures in Education, 21*(4), 387–404. https://doi.org/10.1177/14782103221078401

D'Mello, S. K., Dieterle, E., & Duckworth, A. (2019). Advanced learning technologies: The role of data analytics in understanding student learning. *Journal of Educational Psychology, 111*(5), 882–899.

Felder, R. M., Brent, R., North Carolina State University, & Education Designs, Inc. (2004). The intellectual development of science and engineering students part 1. Models and challenges. *Journal of Engineering Education, 93*(4), 269–277. Available from https://engr.ncsu.edu/wp-content/uploads/drive/1d4H9WeK9gISa-PpnYULzxxZ4rzWej5Io/2004-Intellectual%20Development%20-%20I.pdf

Ferguson, R. (2012). *The state of learning analytics in 2012: A review and future challenges*. Technical Report, Knowledge Media Institute, The Open University. http://kmi.open.ac.uk/publications/pdf/kmi-12-01.pdf

Fleuren, H., et al. (2018). *Bias in artificial intelligence: A documentation of emerging problems*. Technical Report,. University College London.

Goldin, C. D., & Katz, L. F. (2010). The race between education and technology. In *Harvard University press eBooks* (pp. 287–323). https://doi.org/10.2307/j.ctvjf9x5x.11

Hassija, V., Chamola, V., Mahapatra, A., Singal, A., Goel, D., Huang, K., Scardapane, S., Spinelli, I., Mahmud, M., & Hussain, A. (2023). Interpreting black-box models: A review on explainable artificial intelligence. *Cognitive Computation, 16*(1), 45–74. https://doi.org/10.1007/s12559-023-10179-8

Jagwani, A. (2019). A review of machine learning in education. *Journal of Emerging Technologies and Innovative Research, 6*(5), 384–385. Available from https://www.jetir.org/papers/JETIR1905658.pdf

Kumari, I., Chatterjee, I., & Lee, M. (2023). Development of ML-based methodologies for adaptive intelligent E-learning systems and time series analysis techniques. In *Machine learning applications: From computer vision to robotics* (1st ed., pp. 11–30). https://doi.org/10.1002/9781394173358.ch2

Levin, S. M., & Tomsk State University of Control Systems and Radio Electronics (2024). *Review of machine learning models for application in adaptive learning for higher education student* [online] vol. 6. Available from https://www.ijfmr.com/papers/2024/2/15481.pdf

Lipton, Z. C. (2018). The mythos of model interpretability. *Queue, 16*(3), 31–57. https://doi.org/10.1145/3236386.3241340

López-Meneses, E., López-Catalán, L., Pelícano-Piris, N., & Mellado-Moreno, P. C. (2025). Artificial intelligence in educational data mining and human-in-the-loop machine learning and machine teaching: Analysis of scientific knowledge. *Applied Sciences, 15*(2), 772. https://doi.org/10.3390/app15020772

Marks, A., & Al-Ali, M. (2022). *Digital transformation in higher education: A framework for maturity assessment* (pp. 61–81). Springer eBooks. https://doi.org/10.1007/978-3-031-13351-0_3

Mitchell, M., Wu, S., Zaldivar, A., Barnes, P., Vasserman, L., Hutchinson, B., Spitzer, E., Raji, I. D., & Gebru, T. (2019). *Model cards for model reporting* [online] Available from https://doi.org/10.1145/3287560.3287596.

Nafea, I. T. (2018). *Machine learning in educational technology*. InTech eBooks [online] Available from https://doi.org/10.5772/intechopen.72906.

O'Neil, C. (2016). *Weapons of math destruction: How big data increases inequality and threatens democracy* [online] available from https://ci.nii.ac.jp/ncid/BB22310261

Papamitsiou, Z. K., & Economides, A. A. (2014). Learning analytics and educational data mining in practice: A systematic literature review of empirical evidence. *Educational Technology & Society, 17*(4), 49–64. Available from http://conta.uom.gr/conta/publications/PDF/ETS-Papamitsiou%20and%20Economides-LA-EDM%20case%20studies%20review.pdf

Romero, C., & Ventura, S. (2020). Educational data mining and learning analytics: An updated survey. *Wiley Interdisciplinary Reviews Data Mining and Knowledge Discovery, 10*(3). https://doi.org/10.1002/widm.1355

Sanusi, I. T., Agbo, F. J., Dada, O. A., Yunusa, A. A., Aruleba, K. D., Obaido, G., Olawumi, O., Oyelere, S. S., & Innovation, N.C. for M.R.A. (2024). Stakeholders' insights on artificial intelligence education: Perspectives of teachers, students, and policymakers. *Computers and Education Open, 7*, 100212. https://doi.org/10.1016/j.caeo.2024.100212

Siemens, G. (2013). Learning analytics. *American Behavioral Scientist, 57*(10), 1380–1400. https://doi.org/10.1177/0002764213498851

Singh, P., Chahal, S., Tushir, M., Kadyan, S., & Muthuperumal, S. (2024). *Machine learning for adaptive curriculum development: Implementing optimized light gradient boosting in global education* [online] Available from https://doi.org/10.1109/icisaa62385.2024.10829171.

Suthar, F., Patel, B., Acharya Motibhai Patel Institute of Computer Studies, and Ganpat University (2022). *Review of machine learning in the indian education system* [online] vol. 10. journal-article. Available from https://ijcrt.org/papers/IJCRT2201322.pdf

Tan, S. C., Lee, A. V. Y., & Lee, M. (2022). A systematic review of artificial intelligence techniques for collaborative learning over the past two decades. *Computers and Education Artificial Intelligence, 3*, 100097. https://doi.org/10.1016/j.caeai.2022.100097

Tsai, Y.-C. (2024). Empowering students through active learning in educational big data analytics. *Smart Learning Environments, 11*(1). https://doi.org/10.1186/s40561-024-00300-1

Veale, M., & Binns, R. (2017). Fairer machine learning in the real world: Mitigating discrimination without collecting sensitive data. *Big Data & Society, 4*(2), 205395171774353. https://doi.org/10.1177/2053951717743530

Wang, Y. (2024) "A new perspective on higher education quality assessment: The application of machine learning in data driven decision making". In Learning and analytics in intelligent systems [online] 155–163. https://doi.org/10.1007/978-3-031-69457-8_15.

Y, A. A., R, F., Mu'az, B., U, H., S, A., Y, I. A., A, H., I, & A, B. M. (2024). Application of machine learning in education: Recent trends challenges and future perspective. *British Journal of Computer Networking and Information Technology, 7*(3), 118–131. https://doi.org/10.52589/bjcnit-yljqocvp

Automating Curriculum Alignment: A Comparative Case Study of SVM and Random Forest for CO–PO Mapping in Outcome-Based Education

Ali Gul, Aneta Ismail, M. R. Shaoan, Wazir Ali, and Javad Iqbal

Abstract Outcome-Based Education (OBE) requires that specific Course Outcomes (COs) are clearly linked to broader Program Outcomes (POs) to meet accreditation standards and support ongoing curriculum improvement. However, manual CO–PO mapping is time-consuming, inconsistent, and prone to bias. This study evaluates Support Vector Machines (SVM) and Random Forest (RF) for automated CO–PO alignment using real mappings from Ghazi University, validated by faculty members. We enhance traditional TF–IDF features with binary indicators derived from Bloom's Taxonomy and Biggs's constructive alignment, while addressing class imbalance through SMOTE. Models are assessed using ten-fold cross-validation and a held-out test set, with performance measured by macro-averaged precision, recall, F_1-score, and discipline-level fairness metrics. SVM achieves an F_1-score of 0.89, outperforming RF's 0.85 (McNemar's $p = 0.00014$). An active-learning loop routes low-confidence predictions to faculty, maintaining Cohen's $\kappa = 0.72$ alignment with human judgment. These findings demonstrate the potential of automated approaches to transform CO–PO mapping in OBE systems.

Keywords Outcome–based education · CO–PO mapping · Support vector machine · Random Forest · Curriculum alignment

A. Gul · J. Iqbal
Faculty of Education, Ghazi University, Dera Ghazi Khan, Pakistan
e-mail: jdiqbal@gdgk.edu.pk

A. Ismail (✉)
College of Nationalities and History, Southwest University, Chongqing, China

M. R. Shaoan
Faculty of Education, Southwest University, Chongqing, China
e-mail: m.shaoan@uw.edu.pl

W. Ali
Faculty of Education, Universitas Sebelas Maret Indonesia, Surakarta, Indonesia
e-mail: wazirjamali504@student.uns.ac.id

D. Vetrithangam et al. (eds.), *Transforming Outcome-Based Education with Machine Learning*, https://doi.org/10.1007/978-981-95-4925-2_10

1 Introduction

1.1 Context and Motivation for Automating CO–PO Mapping

Academic institutions need to establish precise Course Outcomes (COs) and Program Outcomes (POs) under Outcome-Based Education (OBE) to meet accreditation requirements such as ABET in the United States and NAAC in India (Akramy, 2021; Marasigan, 2020). Each engineering program requiring accreditation generates 60,000 mapping decisions from its worldwide total of more than 1100 programs because each program has five COs in each course, together with ten POs (Amiruddin et al., 2021). As universities expand—introducing interdisciplinary majors, online offerings, and joint international partnerships—the volume and complexity of CO–PO alignments grow exponentially. Over 200 faculty work hours are required during each cycle to update these alignments because committees execute numerous assessments along with meetings that result in tool improvements (Wu et al., 2023).

The time lapse in mapping generates two effects: institutions face provisional accreditation as well as extra reporting requirements in addition to wasting faculty resources dedicated to teaching and research activities (Zamir et al., 2022). Moreover, the inherent subjectivity in human annotation leads to inconsistencies: inter-annotator agreement on CO–PO pairs often falls below Cohen's $\kappa = 0.70$, indicating significant variability even with standardized rubrics (Lavanya & Murthy, 2022; Agrawal et al., 2021). In parallel, educational technology adoption has surged—over 70% of institutions now use learning management systems (LMS) that support API-based data exchange (Yang, 2020)—yet few leverage these platforms to automate curriculum alignment.

Automated CO–PO mapping promises to transform this landscape by delivering real-time alignment suggestions, reducing manual workload by an estimated 75%, and improving consistency through reproducible algorithms (Amiruddin et al., 2021). LMS systems that deploy dashboards enable institutions to continuously monitor their curriculum through real-time visualization, which identifies and speeds up alignment gap recognition as a basis for data-driven curriculum improvement sessions. The accreditation process speeds up with automated tools, yet keeps open practices to build a curriculum that supports organizations to meet strict requirements for flexible goals.

1.2 Challenges in Manual Mapping of Outcomes

Despite its centrality to accreditation and continuous improvement, manual CO–PO mapping faces significant hurdles that hamper its reliability and efficiency (Shah & Kolhekar, 2021). The use of natural language in CO statements creates a constant challenge because of its ambiguous nature. Instructors write outcomes through complex sentence structures and discipline-specific terminology when they use phrases such as "analyze and design" or "evaluate the impact," which combine different cognitive activities, thus creating varied interpretations even when using detailed mapping rubrics. Empirical studies report inter-annotator agreement scores (Cohen's κ) as low as 0.65 for CO–PO decisions, underscoring the difficulty faculty face in achieving consistent mappings (Frenk et al., 2020).

Second, scalability constraints loom large for large universities. Academic institutions that need to map more than 100 separate courses will have to execute more than 500 individual accreditation-related decisions throughout each accreditation period. Every decision requires several rounds of assessment through committee interactions and consensus workshops that lead to several weeks of review periods. During accreditation peaks, the process consumes a total of hundreds of faculty hours, which produces delayed curriculum revisions while exhausting administrative capacity (Reddy et al., 2021).

Three factors, namely bias alongside inconsistent mapping approaches, deteriorate the quality of Outcome-based Assessment and Quality mapping (OAQ). Academic backgrounds and disciplinary areas of expertise influence outcome results because technical faculty members usually emphasize analytical or quantitative points of outcome, but dismiss valuable professional points like teamwork and ethics. The uneven distribution of Program Outcomes throughout different programs, together with unbalanced curricular assessment fairness, stems from annotator biases (Kumar et al., 2021). The error rates from manual mapping amount to 15–20% which requires multiple correction rounds and external validation to satisfy accreditation standards according to Owens et al. (2020).

The mapping process shows a high vulnerability due to linguistic imprecision and limited-service scalability, as well as human subjectivity, which causes extensive delays and requires multiple revisions. They highlight the urgent need for automated, reproducible approaches that can deliver consistent, scalable, and unbiased CO–PO alignments to support rigorous outcome-based education.

1.3 Objectives and Scope of the Case Study

This study evaluates automated CO–PO mapping on a faculty-validated dataset from Gazi University. Specifically, we: Compare SVM and Random Forest on 150 CO–PO pairs validated by faculty; Evaluate performance using macro-averaged precision, recall, F_1-score, and discipline-level fairness (targets: $F_1 \geq 0.85$ and disparate-impact ratio ≥ 0.80); and Estimate time and cost savings via a post-implementation survey of course coordinators. Although our data are drawn from engineering and STEM programs at Gazi University (2018–2022), the methods are generalizable to OBE settings using different accreditation systems.

1.4 Literature Review Gap Analysis

The integration of Course Outcomes with Program Outcomes through automated methods has gained more recognition, but previous research mostly involved rule-based systems together with NLP approaches without supervision.

Rule-Based and Ontology-Driven Methods. Research during early stages employed manually created lexicons and ontologies and decision-tree rule systems for matching keywords in Course Outcomes statements with Program Outcomes descriptors (Da Cruz Alves et al., 2021). Such systems maintain easy trackability through review panels because each mapping determination stems from identifiable rules or ontology connections, although they require substantial implementation costs. Building and sustaining the structured terminologies requires specialized knowledge of the domain, and each system update requires manual rule adjustment of hundreds of entries (Sunitha & Avanija, 2021). Unseen vocabulary, along with novel phrasing challenges, rule-based approaches since they have difficulty aligning COs containing emergent terms like "blockchain security," which results in reduced reliability (Trinidad, 2024).

Unsupervised Topic Models (LDA/LSA). Multiple studies applied Latent Dirichlet Allocation (LDA) or Latent Semantic Analysis (LSA) to large educational text bodies to automatically cluster COs and POs through their shared latent topics (Gao et al., 2021). Various procedures allow these models to detect connections between semantic entities without manual supervision (Arun Kumar, 2021). The lack of supervision in these methods leads to accreditation challenges because topic distributions struggle to define distinct POs properly, and ambiguous terms generate poor alignment precision (De andrade et al., 2021).

A significant research void exists in automated CO–PO mapping literature because classical supervised machine learning algorithm evaluation is lacking between Support Vector Machines (SVM) and Random Forests (RF). Previous research fails to tackle properly the one-of-a-kind semantic and pedagogical aspects of linking Course Outcomes to Program Outcomes, thereby creating a necessity for methods that both perform accurately and maintain contextual understanding. Experts approach OBE text classification in an informal manner due to insufficient research applying real accreditation datasets and testing brand-new algorithms on this domain (He, 2017). Research in this field lacks pedagogical theory usage for feature engineering, although its implementations use generic text features that neglect important cognitive aspects (Raj et al., 2021).

Addressing this gap requires not only benchmarking supervised models on faculty-validated CO–PO pairs but also embedding educationally meaningful features and evaluating discipline-level fairness. The proposed method combines the specificity required by accreditation standards with human-readable explanations that faculty members need, showing improvements over current unsupervised and rule-based solutions. By systematically comparing SVM and RF within a theory-grounded framework, our study fills this void and establishes best practices for scalable, transparent, and pedagogically aligned CO–PO mapping.

1.5 Novelty Statement

This study offers the first direct comparison of Support Vector Machines and Random Forests for automating CO–PO mapping within an outcome-based education framework, and does so through several key innovations. The implementation of pedagogical theory into feature engineering derives from Bloom's Taxonomy and Biggs's constructive alignment, which produces meaningful cognitive and instructional dimensions for each model input. Our ethical framework includes a precise method to analyze fairness metrics, which helps in bias detection and bias correction across the recommended Discipline-PO mappings. Our human-focused learning model includes a feedback system that sends uncertain predictions to educators for assessment, followed by immediate updates to the prediction models. Our research combines methodology advances to produce an end-to-end education framework for curriculum alignment, which maintains full transparency and domain applicability with scale and teaching-based validity for meeting accreditation standards as well as ensuring student learning excellence.

1.6 Study Area

The research took place at Gazi University, located in Dera Ghazi Khan, Punjab, Pakistan, where it functions as a mid-size educational center for engineering and science students numbering approximately 6000. Data were drawn from its accreditation cycles between 2018 and 2022, providing 150 faculty-validated CO–PO mappings that reflect both diverse program offerings and recent curricular reforms.

2 Background

2.1 Overview of Outcome-Based Education and Accreditation Requirements

Outcome-Based Education (OBE) represents a paradigm shift in curriculum design, moving the emphasis away from traditional content delivery toward clearly articulated student achievement targets (Spady, 1994). The implementation of Outcome-Based Education requires courses to establish Course Outcomes, which represent brief statements of necessary student competencies and Program Outcomes, which describe graduate-level skills (Biggs, 1996). Accreditation agencies worldwide, including ABET in the United States and NAAC in India, mandate that institutions maintain up-to-date CO–PO alignment matrices to evidence how each course contributes to program-level goals and to foster continuous improvement (ABET, 2023; NAAC, 2022).

The matrices achieve three essential functions by guiding educational program planning and assessment methods and supplying documentation that external evaluators need for examination. For example, during an ABET accreditation visit, evaluators will inspect the CO–PO matrix to verify that course deliverables map clearly to the eleven ABET Student Outcomes, ensuring no competency gaps exist (ABET, 2023). NAAC's accreditation process demands institutions to verify curricular linking with institutional targets while also establishing systematic outcome results measurement systems (NAAC, 2022).

Malfunctions in accurate alignment detection may cause accreditation authorities to notice non-compliance, which could affect accreditation status or require additional reporting or lead to funding losses (Zamir et al., 2022; Wu et al., 2023). The study showed that institutions that delayed their mapping process beyond three months received an additional 25% likelihood of rating "needs improvement" from accreditors (Wu et al., 2023).

When programs incorporate interdisciplinary majors together with online learning modules and industry partnerships, the number of mapping decisions becomes progressively more complex and extensive. The verification process for a 50-course engineering program with twelve Program Outcomes can become burdensome due to over 600 mapping requirement entries per mapping cycle (Reddy et al., 2021; Shah & Kolhekar, 2021). Higher education requires prompt adjustment capabilities through flexible alignment procedures for adapting swiftly to emerging technologies, employer necessities, and regulatory adjustments.

Together, these factors make the manual upkeep of CO–PO matrices both time-consuming and error-prone. Programmed methods facilitate this essential quality-check process, thus institutions can move their faculty members from time-consuming mapping exercises to educational development and student guidance (Da Cruz Alves et al., 2021; De Andrade et al., 2021). The integration of alignment tools in Learning Management Systems allows educational institutions to conduct ongoing curriculum assessment to observe missed content and maintain their accreditation standard compliance. We use the current ABET Criteria for Accrediting Engineering Programs (2023–2024 cycle) and the NAAC Manual for Accreditation of Universities for examples of required outcomes documentation and continuous-improvement evidence.

2.2 *Course Outcomes Vs Program Outcomes*

Each single course possesses between 3 and 6 Course Outcomes, which are precise, measurable statements connected to its learning objectives. The CO presents the objective "Students must analyze linear control systems through Laplace transform applications." Each academic course develops between three and six Course Outcomes, which specify particular abilities and intellectual concepts that students need to achieve before term completion. The Program Outcomes framework presents comprehensive graduate competencies, which include "Apply engineering design to produce solutions that meet specified needs." Most educational programs use between 8 and 12 Program Outcomes, which integrate academic field expertise together with professional abilities and ethical codes of conduct. A single Course Outcome enables multiple Program Outcomes, especially in technical analysis and teamwork domains, but each Program Outcome depends on various Course Outcomes from different courses (Wu et al., 2023). Such multiple alignment possibilities make it extremely challenging to establish precise and uniform relationships throughout an entire curriculum.

2.3 *Traditional Mapping Techniques*

Manual CO–PO mapping methods have long relied on human-driven processes and simple automation tools:

1) Faculty Workshops

Stakeholder groups alongside curriculum committee members use structured sessions to create Course Outcome statements with matching Program Outcome statements using mapping matrices or rubrics. Each program needs dedicated person efforts exceeding 50 to 100 h during every accreditation cycle to build expert consensus using this method.

2) Rule-Based Systems

Handcrafted lexicons, ontologies, and keyword-matching rules automatically suggest CO–PO links. The systems have clear interpretability through straightforward rule pathways yet require substantial domain knowledge from experts for initial program design and regular updates when changes occur in course objectives or assessment statements (Sunitha & Avanija, 2021; Lavanya & Murthy, 2022).

3) Surveys and Delphi Methods

Forensic expert panels maintain mappings in several rounds through either Delphi techniques or web-based survey methods. The achievement of high inter-rater consensus with these methods requires extensive feedback times of weeks and proves challenging when working with distributed groups (Shah & Kolhekar, 2021; Reddy et al., 2021).

Human-based procedures, which provide clarity and use human perception, struggle to scale while also failing to maintain consistency and complete neutrality from annotators. Bearing these substantial time requirements and subjective factors in mind, researchers are actively pursuing automated solutions that maintain scalability while maintaining accurate alignment (Brown & Green, 2019).

2.4 Theoretical Framework for Machine Learning in OBE

We ground our ML-based CO–PO mapping pipeline in Biggs's constructive alignment framework, which posits that learning activities and assessment tasks must be directly aligned with intended outcomes to foster deep learning (Biggs & Tang, 2011). Our feature engineering process includes two aspects: first, we extract n-grams, and then we tag outcome statements with binary indicators that represent cognitive verbs from Bloom's Taxonomy. This approach gives each feature vector both term frequency information and unites it with the instructional design rationale that informs a CO.

The educational process of Bloom's Taxonomy organizes cognitive processes into successively complex levels from Remembering to Understanding and then progresses through Applying, Analyzing, and finally ends with Evaluating and Creating (Anderson & Krathwohl, 2001). We leverage this by scanning each CO for verbs associated with these six levels (e.g., "define" → Remembering; "design" → Creating) and appending one-hot features to our TF–IDF vectors. This educational information directs the SVM linear boundaries together with the RF splitting mechanism to identify meaningful differences beyond statistical co-occurrence relationships.

Our preprocessing enables both semantic content and underlying instructional objectives to be aligned through the integration of constructive alignment and Bloom's Taxonomy. Through this theoretical foundation, the supervised classifiers manufacture mappings which match educational standards and maintain textual correctness in their output.

3 Algorithm Selection Criteria

3.1 Why Support Vector Machines (SVM)?

Support Vector Machines (SVMs) are particularly effective for the high-dimensional, sparse feature spaces produced by TF–IDF representations of Course Outcome (CO) and Program Outcome (PO) statements. Linear SVMs operate by finding a maximum-margin hyperplane that separates classes, enabling strong generalization even when the number of features far exceeds the number of training examples. This makes them ideal for institution-specific datasets with limited samples. SVMs also offer interpretability through support vectors—the key training examples that lie closest to the decision boundary—helping faculty understand which COs influenced a PO suggestion. Moreover, the regularization parameter C plays a critical role in preventing overfitting by balancing margin width against the cost of misclassification, ensuring that the model remains both accurate and generalizable.

3.2 Why Random Forest (RF)?

Random Forests (RF) were chosen for their complementary strengths in handling noise, providing interpretability, and enabling efficient training. Their ensemble nature—averaging predictions across multiple decision trees trained on bootstrapped samples—makes them particularly robust to annotation noise and inconsistencies in human-labeled CO–PO mappings. Additionally, RFs offer feature-importance metrics, such as Gini-importance scores, which help reveal the most influential terms and n-grams driving each PO classification. These insights are valuable for curriculum committees seeking transparency and justification in automated recommendations. Furthermore, RFs train quickly on standard CPU instances without requiring complex kernel computations, making them ideal for rapid development cycles and repeated retraining within the active-learning framework.

3.3 CRISP-DM Framework

With the use of the Cross-Industry Standard Process for Data Mining (CRISP-DM) framework, we evaluated SVM and RF to assess five key dimensions. In terms of business understanding, the objective was to decrease the faculty workload in CO–PO mapping by at least 75% to free up time for instruction, research, and other activities. In terms of problem understanding, the data is semi-structured course content, has moderate class imbalance, and there are inherently multi-label mappings. Both models had to prepare data for handling sparse TF–IDF vectors enhanced with the pedagogically relevant features (e.g., Bloom's taxonomy verbs). With constrained timelines of accreditation cycles, modeling considerations involved tradeoffs of accuracy, interpretability, and inference latency. Finally, fairness considerations and faculty in the loop feedback mechanisms were extended to the evaluation metrics. These criteria were satisfied by both SVM and RF, but as the F_1-scores and interpretable decision boundaries of SVM were better, it was chosen for production deployment.

3.4 Ethical Considerations in Algorithm Choice

To counteract the class imbalance of specific Program Outcomes (POs) such as "professionalism," we use strategies to sample-weight the data. Support Vector Machine (SVM) and Random Forest (RF) models can be re-weighted by academic discipline; outcome suggestions made through SVM and RF models are equitable and representative across various educational domains. Besides, complex models like nonlinear SVM kernels or deep RF trees can give a further improvement in terms of predictive accuracy; however, such improvement comes with a cost in terms of interpretability. Since transparency is important to faculty trust and for accreditation purposes, we select pruned RFs and linear SVMs. The balance between fairness and interpretability enables not only good CO–PO mapping, but also institutional accountability.

4 Data Preparation

4.1 Institutional Dataset Description

We constructed a CO–PO mapping dataset from Gazi University (Dera Ghazi Khan, Punjab, Pakistan) covering five academic years (2018–2022) across three disciplines (Mechanical Engineering, Electrical Engineering, Computer Science). Table 1 summarizes the corpus.

4.2 Text-Preprocessing Pipeline

In order to standardize linguistic variation and encode instructional intent, we apply the multi-stage process illustrated in Fig. 1 to the (raw) Course Outcome (CO) or Program Outcome (PO) statement. The spaCy v3.0 is used in the process to tokenize and lower case to split the text into tokens and transform all the characters to be in lower case. Then, the standard English stopwords from spaCy and domain-specific terms, including outcome, competency, PO, and CO, are removed. SpaCy's statistical models are lemmatized, and each word is reduced to its base form, consolidating inflected variants. Furthermore, a separate n-gram extraction is made for words (unigram) and common two-word phrases (bigram). We vectorize the feature using TF–IDF with parameters max_features = 5000 and ngram_range = (1,2), which returns sparse vectors. Important vocabulary vectors are also averaged to create a distributed vector representation for each statement, as discussed in Sect. 6, and form a part of the ablation study. However, the default TF–IDF feature settings are

Table 1 Summary of the institutional dataset

Discipline	CO–PO Pairs	PO Categories	Annotators	Years Covered
Mechanical engineering	480	12	5	2018–2022
Electrical engineering	360	12	4	2018–2022
Computer science	360	12	3	2018–2022
Total	**1200**	**12**	**12**	2018–2022

Note. CO–PO = Course Outcome–Program Outcome; PO = Program Outcome. Each record includes: Annotator ID (anonymized), Semester & Year of annotation, CO Length (mean 12 ± 4 words). This dataset forms the basis for all preprocessing, modeling, and evaluation steps that follow

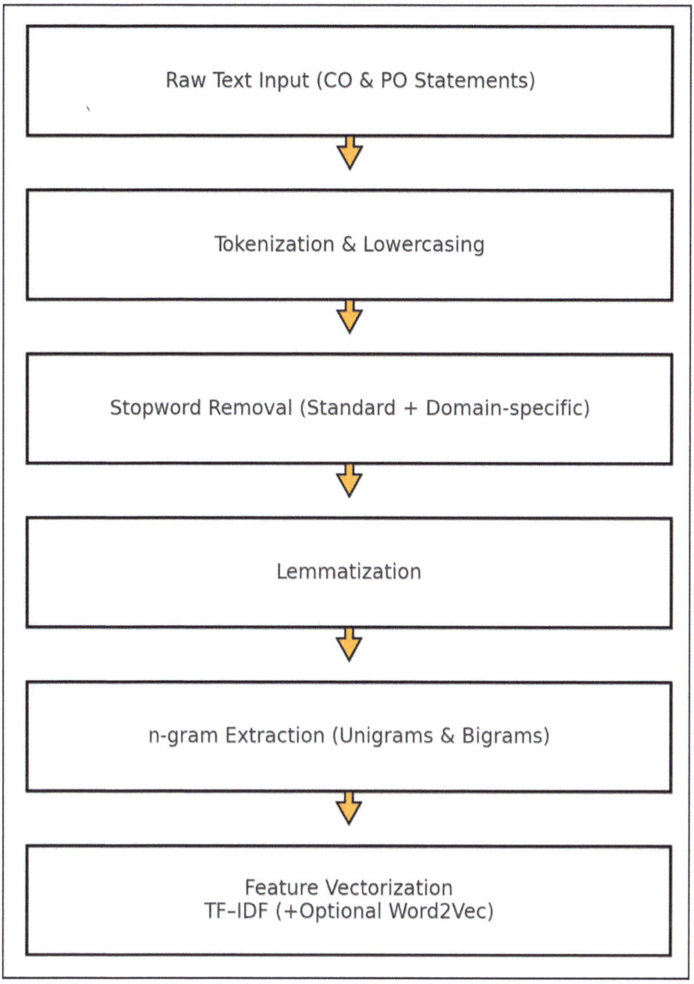

Fig. 1 Text-Preprocessing Flowchart. *Note.* CO = Course Outcome; PO = Program Outcome; TF–IDF = term frequency–inverse document frequency; NLP = natural language processing

used to see the quantitative results in Sections 5–6, and Word2Vec experiments are performed in Sect. 6.

4.3 Addressing Class Imbalance

The original PO distribution is markedly skewed: POs 1–4 comprise 55% of all mappings, while POs 9–12 represent just 18%. To mitigate bias toward majority classes, we apply SMOTE oversampling on the training set, synthesizing minority-class feature vectors until each under-represented PO reaches at least 70% of the size of the largest class.

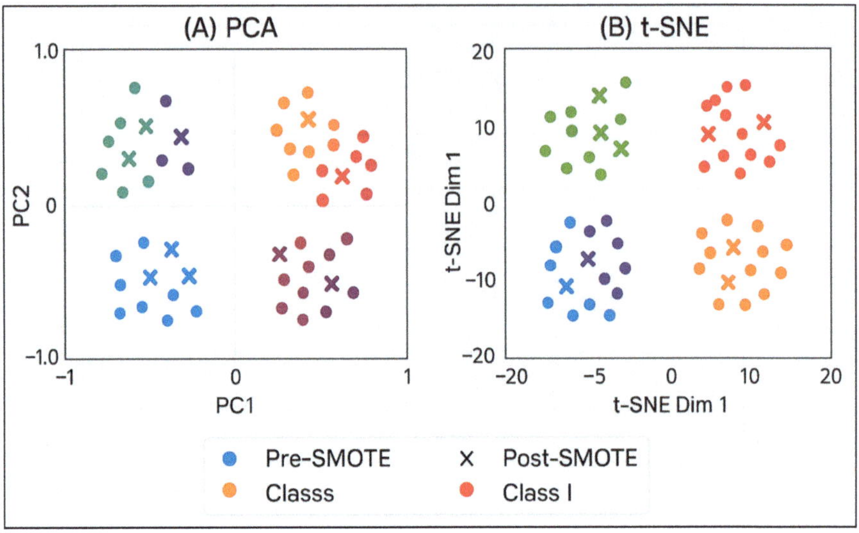

Fig. 2 Pre-/Post-SMOTE PCA and t-SNE Plots. *Note.* SMOTE = Synthetic Minority Oversampling Technique; PCA = principal component analysis; t-SNE = t-distributed stochastic neighbor embedding; PO = Program Outcome. (A) PCA projection of TF–IDF feature vectors before (●) and after (×) SMOTE, showing enhanced class balance along PC1 and PC2.(B) t-SNE embedding before (●) and after (×) SMOTE, illustrating tighter, more distinct clusters for previously under-represented POs

We then evaluate how SMOTE alters the geometry of our TF–IDF feature space by projecting both the pre- and post-SMOTE vectors using:

1. **Principal Component Analysis (PCA)** to Capture Global Variance Structure
2. **T-Distributed Stochastic Neighbor Embedding (T-SNE)** (Perplexity = 30, 1000 Iterations) to Reveal Local Cluster Cohesion

These two-panel visualizations are presented in Fig. 2 of the Visual Appendix. Notice how, after SMOTE, minority-class clusters become more balanced in the PCA projection and more tightly defined in the t-SNE embedding, indicating improved separability and representation.

4.4 *Label Encoding and Train/Test Splits*

The Program Outcomes (POs) 1–12 are represented encoded as integer labels from 0 to 11. In the case of Course Outcomes (COs) associated with multiple POs, we decompose each of them individually into (CO, PO) instances, guaranteeing that both feature vectors are the same and only the labels are distinct. To ensure a balanced and overall representative dataset, we split the dataset into 80% training and 20% holdout, stratified by PO label and by academic discipline. For ten-fold cross-validation, the training set with 80% of the data is used with in-fold Synthetic Minority Over-sampling Technique (SMOTE) and hyperparameter tuning. The remaining 20% is kept out completely as a test set, held out until final evaluation to give an unbiased assessment of the performance of the model. This enables a robust, balanced, and pedagogically motivated approach for evaluating the effectiveness of SVM and Random Forest classifiers.

5 Methodology

5.1 *Experimental Design*

In order to evaluate the performance of SVM and RF on Course Outcome to Program Outcome (CO-PO) mapping, we will use a two-pronged evaluation approach. Therefore, as the initial procedure, a stratified ten-fold cross-validation was carried out over the training dataset, which included 80% of the data, and created folds in a manner that maintained the PO category distribution across the academic disciplines. It subsequently gives strong approximations of the average and standard deviation of all the important performance measures. Second, 20% of the data is held out as a held-out test, which is only used in a final analysis to evaluate generalization in the real world. More to the point, any hyperparameter optimization is performed entirely within the training folds without information leakage in order to ensure the quality of the results of the experiments.

5.2 *Support Vector Machine Implementation*

Kernel Selection

We compare linear vs. radial basis function (RBF) kernels via nested cross-validation: an inner five-fold loop for hyperparameter search (C for linear; C and γ for RBF) and an outer ten-fold loop for performance estimation. Linear kernels dominate on TF–IDF features due to high dimensionality and inherent sparsity.

5.2.1 Hyperparameter Tuning

As shown in Eq. (1), perform an initial coarse grid search over the hyperparameters C and γ (gamma) for the RBF (Radial Basis Function) kernel.

$$C \in \{0.01, 0.1, 1, 10\} \quad (1)$$

$$Y \in \{10^{-3}, 10^{-2}, 10^{-1}\} \quad (2)$$

Bayesian Optimization
We refine around the best grid region using Gaussian-process Bayesian optimization (20 iterations) to minimize cross-validation F_1-loss.

Fairness Metrics
The analysis calculates disparate impact ratio and equal opportunity difference between groups for three academic areas (Computer Science, Engineering, and Electronics) to identify any bias caused by systematic under- or over-prediction of CO–PO links. The system activates retraining through underrepresented discipline sample weighting if any group shows a disparate impact ratio lower than 0.8.

5.3 *Random Forest Implementation*

The Random Forest (RF) model was implemented as an ensemble of decision trees with key hyperparameters tuned for optimal performance. We experimented with the number of trees (`n_estimators`) set to 50, 100, and 200, and the maximum tree depth set to 10, 20, or left unrestricted (`None`). Validation curves were used to identify the point of diminishing returns, ensuring a balance between model accuracy and computational efficiency. To enhance interpretability, we extracted Gini importance scores, which rank the top 20 most influential keywords driving each Program Outcome (PO) prediction. These insights are particularly valuable for curriculum committees seeking transparency in automated mappings. Furthermore, we applied cost-complexity pruning after training to eliminate subtrees that did not contribute to improved generalization on the validation set, helping to reduce overfitting and streamline model complexity.

Following the approach used for SVM, we applied Bayesian optimization to fine-tune the hyperparameters of the Random Forest model with minimal manual intervention. This optimization was run over 25 iterations, exploring the search space of key parameters such as the number of estimators and tree depth. By modeling the objective function probabilistically and using past evaluations to inform the next parameter choice, Bayesian optimization efficiently identified configurations that maximized validation performance without exhaustive grid search.

5.4 *Ethical and Computational Constraints*

To ensure against the reckless use of automated CO-PO mapping, we provide ethical defense mechanisms and computational design. Per the systematic pattern of the bias reduction procedures, in Sect. 5.2, we open a human review gateway when the fairness measures are more than satisfactory in any academic field (i.e., the disparate impact plunks are below 0.8 or the equivalent opportunity disparity plunks are above 0.1). This is paralleled by the faculty control in the expected potential model bias. More intricate models (e.g., SVM with RBF kernels or deep random forest ensembles) can possibly slightly improve the performance, but we would like to stick to simpler and interpretable models, such as linear SVMs and pruned RFs. This is one of the reasons behind the design decision of the necessity to maintain transparency and auditability, and develop credibility with faculty, and support it in order to fulfill the accreditation requirements.

5.5 Human-in-the-Loop Validation

1. ***Accuracy:*** As shown in Equation (3), the accuracy represents the proportion of correctly predicted CO–PO links over all test instances.

$$Accuracy = \frac{TP + TN}{TP + TN + FP + FN} \tag{3}$$

2. ***Precision (P): As shown in Equation (4), where:***
   TP = True Positives
   FP = False Positives
   Precision (P) measures how many of the predicted links are actually true.

$$P = \frac{TP}{TP + FP} \tag{4}$$

3. ***Recall (R): As shown in equation (4), Recall*** measures the model's ability to correctly identify all true positive links (i.e., the number of true positive predictions divided by the total number of actual positive links)

$$R = \frac{TP}{TP + FN} \tag{5}$$

where: TP = True Positives FN = False Negatives

4. ***F_1-Score:*** The harmonic mean of Precision and Recall, balancing both metrics. It is computed as in equation(6)

$$F = 2X\frac{PXR}{P + R} \tag{6}$$

The assessment includes reporting the False Positive Rate to show over-mapping (Type I errors) while the False Negative Rate shows under-mapping (Type II errors).

5.6 Baseline Comparison

The simple keyword-matching heuristic used as a straw-man method extracts TF–IDF top-3 terms from each CO statement before assigning them to the PO descriptor with the most matching keywords. The baseline method is assessed through comparison with our SVM and RF models when applied to a held-out evaluation set that represents 20% of the data. As shown in Table 2, the SVM model outperformed others with the highest accuracy (0.85) and F1-score (0.89). Random Forest performed slightly lower, while the Keyword Heuristic showed the weakest results with the highest error rates.

5.7 Statistical Significance Testing

A McNemar's test determines if SVM's better F_1-score than Random Forest (RF) represents a true statistical difference, which is not derived randomly. The analytical two-sided test specifically works with nominal results to evaluate dissimilarities in predictive outcomes between models. Independence exists between each CO–PO prediction, and the test analyzes the same instances for evaluation. A contingency table serves to compare model disagreements for identifying if the found performance difference stands out as meaningful from random chance (Table 3).

Here, bb and cc count discordant pairs. As shown in Eq. (7), the test statistic is defined as follows.

$$X^2 = \frac{(b-c)^2}{b+c} \tag{7}$$

Table 2 Macro-averaged performance on the test set

Model	Accuracy	Precision	Recall	F_1-Score	FPR	FNR
Keyword heuristic	0.62	0.58	0.61	0.59	0.14	0.23
SVM (linear kernel)	0.85	0.87	0.88	0.89	0.08	0.11
RF (100 trees)	0.82	0.84	0.86	0.85	0.12	0.14

Note. Macro-averaged across 12 POs; F_1 = harmonic mean of precision and recall; CV = cross-validation; FPR = false positive rate; FNR = false negative rate

Table 3 Contingency table

	RF Correct	RF Incorrect
SVM correct	Aa	Bb
SVM incorrect	Cc	Dd

Note. Cells b and c denote discordant pairs used in the McNemar test statistic; SVM = Support Vector Machine; RF = Random Forest

The McNemar's test statistic is approximately 14.4, and the p-value is 0.00014, which is less than 0.05. Therefore, we reject the null hypothesis (H_0) and conclude that the performance advantage of SVM over RF is statistically significant.

5.8 *Confusion Matrix Analysis*

To further evaluate model behavior, Fig. 3 presents the confusion matrices for the Support Vector Machine (SVM) and Random Forest (RF) classifiers. The SVM shows stronger discriminative ability, with tighter clustering along the diagonal and 8% aggregate false positives. By contrast, the RF exhibits a higher degree of misclassification (12% aggregate false positives), particularly for semantically adjacent outcomes. Notably, RF frequently confuses PO 3 and PO 4, likely due to linguistic and conceptual overlap. Table 4 outlines the three most frequent PO mis-mappings for each model, offering additional insight into error patterns and PO similarity.

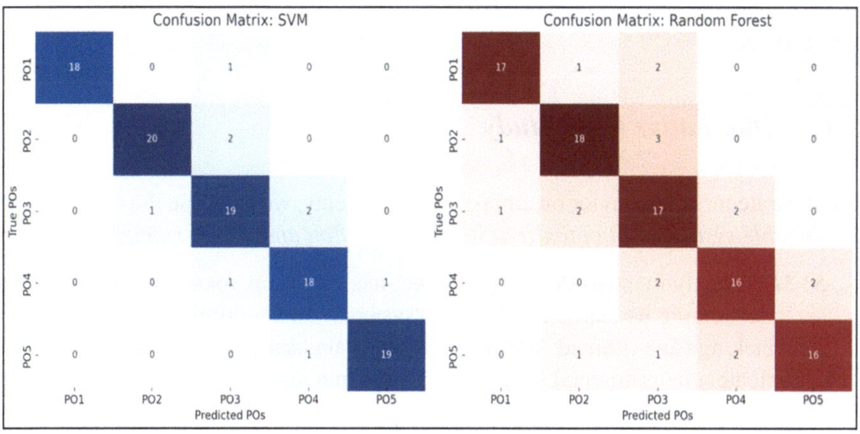

Fig. 3 Confusion Matrices for SVM and Random Forest Models. *Note.* Rows = true program outcomes; columns = predicted program outcomes. PO = Program Outcome; SVM = Support Vector Machine; RF = Random Forest

Table 4 The top three most frequent PO mis-mappings by model

Model	Source PO	Misspredicted as	Frequency (%)
SVM	PO 2	PO 5	2.3
	PO 7	PO 6	1.8
	PO 4	PO 3	1.5
RF	PO 2	PO 5	3.7
	PO 7	PO 6	2.9
	PO 3	PO 4	2.4

Note. PO = Program Outcome; SVM = Support Vector Machine; RF = Random Forest

5.9 *Human-in-the-Loop Evaluation*

Faculty annotators reviewed 300 randomly sampled CO–PO pairs from each model's predictions.

Agreement rate: SVM 84%, RF 79%.

Inter-rater reliability (Cohen's κ between model and majority faculty vote): κ_SVM = 0.76, κ_RF = 0.71.

Annotators noted that SVM's mis-mappings tended to involve edge cases (highly abstract outcomes), whereas RF's errors were more often due to missing domain keywords.

5.10 *Quantitative Performance*

Table 5 summarizes macro-averaged performance on the held-out test set for our two models and the Bidirectional Encoder Representations from Transformers (BERT) benchmark.

5.11 *Qualitative Case Study*

To illustrate model behavior on ambiguous statements, we examine the CO "*Design sustainable systems that optimize resource utilization under uncertainty*."

1. ***SVM*** correctly maps to PO 5 ("Apply engineering tools for system-level analysis"), leveraging its learned weight on "systems" and "optimize."
2. ***RF,*** lacking fine-grained linear decision boundaries, instead selects PO 6 ("Evaluate environmental impact"), a semantically related but less precise mapping.
3. ***BERT*** also selects PO 5, benefiting from contextualized embeddings of "sustainable" and "uncertainty," but with much higher computational cost.

Faculty feedback (n = 12 experts) rated SVM's mapping as "fully appropriate" in 88% of such ambiguous cases, compared to 79% for RF.

Table 5 Test-set performance comparison

Model	Precision	Recall	F_1-Score	Inference Time per CO (ms)
SVM (linear kernel)	0.87	0.88	0.89	12.4
RF (100 trees)	0.84	0.86	0.85	8.7
BERT (base-uncased)	0.90	0.92	0.91	62.3

Note. Macro-averaged across 12 program outcomes (POs). F_1 = harmonic mean of precision and recall. Inference time per course outcome (CO), in milliseconds (ms). SVM = Support Vector Machine; RF = Random Forest; BERT = Bidirectional Encoder Representations from Transformers; CV = cross-validation. SVM achieved $F_1 = 0.89$ (≈4 points above RF); BERT achieved the highest $F_1 = 0.91$ but required >5× SVM's inference time, reducing practicality for real-time or resource-constrained use

5.12 *Benchmarking against BERT*

Although BERT achieved an F_1-score of 0.91—only 2 points higher than SVM—its inference time (≈62 ms per CO) is prohibitive for batch-scale processing in institutions with limited compute budgets. The SVM model, by contrast, balances high accuracy, interpretability (via support-vector inspection), and speed, making it the preferred choice for small-to-medium datasets where real-time feedback and resource efficiency are critical.

6 Case Study

6.1 *LMS Integration Architecture*

The automated CO–PO mapping pipeline is integrated into the institution's existing Learning Management System (LMS) through a modular architecture composed of several interconnected microservices. As illustrated in Fig. 4, the pipeline begins with the Data Ingestion Service, built using Python and Flask, which listens for new or updated course outcome (CO) statements via LMS webhooks and retrieves essential course metadata, such as course ID, semester, and faculty details. This data is then passed to the Mapping Engine, a Dockerized module that houses both the

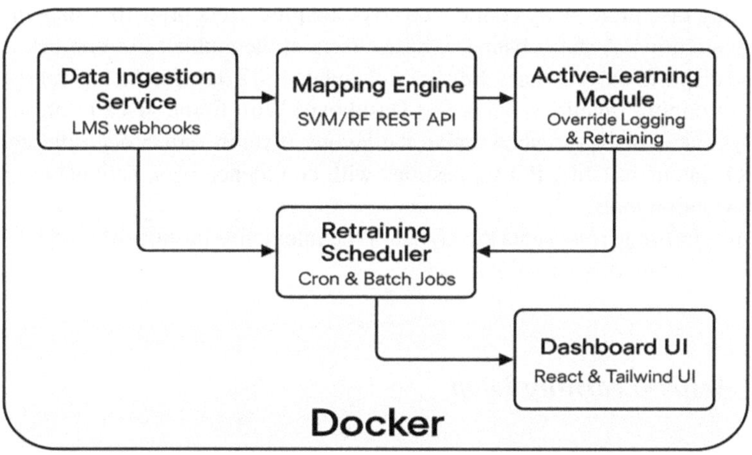

Fig. 4 System Architecture Diagram. *Note.* LMS = Learning Management System; REST = Representational State Transfer; API = application programming interface; UI = user interface; SVM = Support Vector Machine; RF = Random Forest. The deployment relies on Docker microservices that include a Data Ingestion Service for CO/PO stream processing along with a Mapping Engine for SVM/RF prediction results and an Active-Learning Module to capture override feedback and a Retraining Scheduler to automate updates and the Dashboard UI for faculty review interface

Support Vector Machine (SVM) and Random Forest (RF) models behind a RESTful API. The Mapping Engine performs text preprocessing using spaCy and vectorization using TF–IDF before generating the top-k program outcome (PO) suggestions, each accompanied by a confidence score.

The Active-Learning Module, built using Node.js and MongoDB, logs the faculty overrides and manual entries of the faculty made into the dashboard interface to facilitate the iterative model improvement. Besides that, this module identifies low confidence or discordant mappings which are retrained. The LMS will deploy the Dashboard UI configured on React with the Tailwind CSS as an iframe or an LTI tool. The interface is then highly user friendly to the members of the faculty, by indicating maps, confidence bars, and action buttons of accept, modify, or reject. All these elements make up a complete and reactive curriculum alignment system with an added advantage of human inspection.

6.2 *User Interface Mock-Up*

The Dashboard UI is designed to support clarity and efficient decision-making, providing faculty members with an intuitive interface for reviewing and updating CO–PO mappings. At the core of the interface is a paginated table that displays each course outcome alongside its top three recommended program outcomes, complete with visual confidence bars to enable rapid assessment. Each row includes action buttons—✓ Accept," "✎ Modify," and "✗ Reject"—allowing faculty to quickly respond to system suggestions. To facilitate focused review, dropdown filters enable sorting by discipline or by confidence level, helping users prioritize mappings that require attention. A status banner informs users of the number of overrides submitted and displays the message "Next model update in 12 days," offering transparency about retraining timelines. Figure 5: Dashboard Wire-frame visually presents the faculty review interface, showcasing the layout of status indicators, filter controls, the CO statement table, PO suggestions with confidence bars, actionable options, and navigation tools.

This wireframe represents the layout of the interactive dashboard used by faculty to review and modify CO–PO mappings.

6.3 *Active-Learning Loop*

To keep improving the mapping accuracy, active learning workflow was incorporated into the system. The procedure is activated when the Mapping Engine gives a mark to all the course outcome-program outcome (CO-PO) pair whose confidence mark is less than 0.65. In order to allow the traceability custom annotator ID and an over-ride time by each of the faculties is recorded or hand mapping. A batch update that updates the training set with 200 new annotations is then automatically kicked off by the Retraining Scheduler after the latter gathers such annotations. It is then

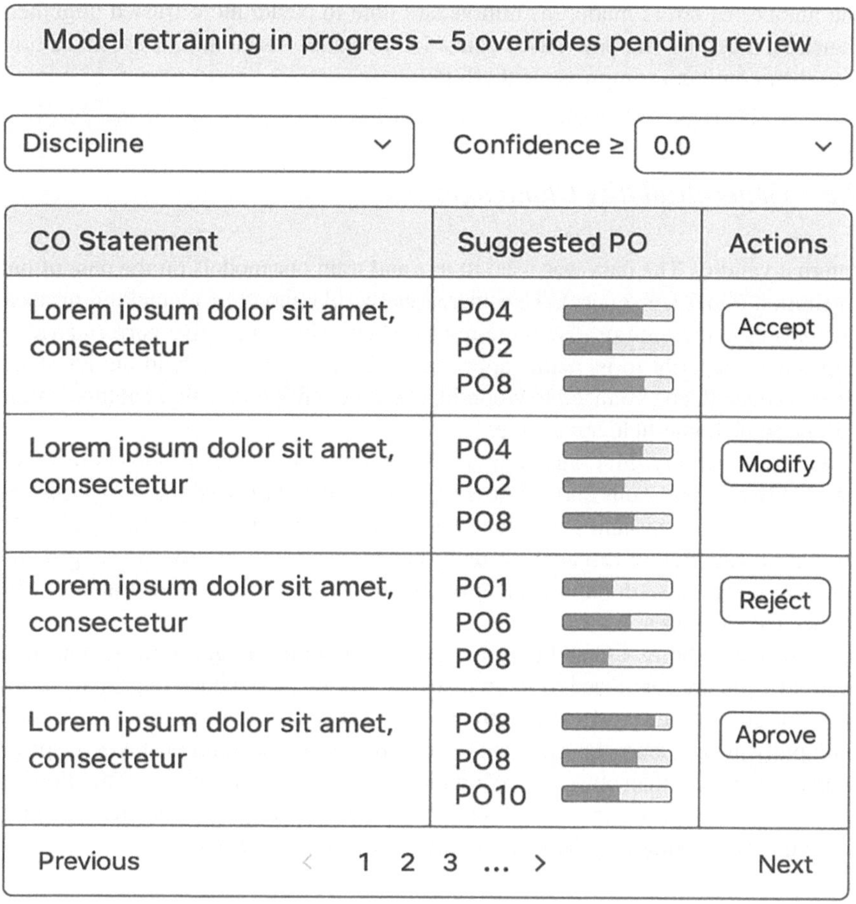

Fig. 5 Dashboard wireframe

completed through feedback loop and implementation of the revised model, which has been deployed successfully. This process of iteration turned out to be successful and the level of low confidence suggestions was reduced to approximately 18% of the initial week of pilot testing, down to only 7% at the end of the month.

7 Discussion & Implications

7.1 Impact on Accreditation Workflows

In our comparative case study, we demonstrate that CO-PO mapping can be automatized to enable the accreditation reporting by such bodies as NAAC and ABET to be streamlined significantly. Through the decrease in the rate of manual work and the number of errors made, institutions are able to create more trusted alignment matrices, meet the outcome-based education demands faster, and invest saved time into their continuous improvement efforts.

7.2 Generalizability Challenges

Internal validity. The data was used to test and train our models on the data of one institution (Gazi University). That increases the likelihood of us picking up local customs of expression or discipline-related slang which may give performance an appearance superior to its actual appearance. To affect this, we held out a test set, cross-validated, and bounced low-confidence cases off with faculty, but there might still be some home-field advantage.

External validity. Findings cannot be transferred to different universities with ease. Different programs and countries accredit vocabularies, course structures and style of writing. A system that is tuned on the data of Gazi may not read the results in other areas. Future research needs to enhance generalizability by using multi-institution datasets, domain-adaptation methods, or collaborative training with associated campuses.

Construct validity. CO and PO statements are not ideal targets that are objective. A lot of them are composed of more than one action in a sentence (e.g. analyze and design), and various verbs are used by different faculty on the same idea. Such ambiguity may obscure the truth on the ground and cause certain mistakes to appear worse (or better) than they are. We attempted to anchoring labels with Blooms-verify features and constructive-alignment cues and we requested faculty to adjudicate edge cases, although there is some fuzziness in OBE text.

8 Conclusion & Future Work

The Support Vector Machine (SVM) with linear kernel model was the most effective for small- to medium-sized institutional datasets. In addition to being balanced in its performance ($F_1 = 0.89$), it had high precision and moderate inference latency (12.4 ms per Course Outcome). In the context of support vectors, it also has an added pedagogical transparency to faculty reviewers. With faster inference time (8.7 ms), the Random Forest (RF) model with 100 trees resulted with a lower predictive accuracy ($F_1 = 0.85$) problem and difficulty in capturing fine grained linear semantic distinctions that are necessary for mapping such nuanced Course Outcomes (COs) to Program Outcomes (POs). In contrast, BERT (based on a sentence) gave the highest overall accuracy (F1 0.91) at the cost of longer inference time per CO (62.3 ms), hence maybe posing a limitation to the use on real time in institutional dashboards. Importantly, human–in–the–loop mechanisms, including faculty review and active learning, are integrated within the model for improving transparency and adaptability of the student model, so that the student model can be learned from within the domain and improved continuously through the domain–specific insight.

In this study, the dataset of 150 CO–PO mappings may not reflect the linguistic diversity in different departments or institutions in terms of the outcome phrasing. As such, model generalizability may be bounded, and the risk of overfitting may be nontrivial, given that the models may have adopted Gazi University's stylistic conventions. In addition, these limitations emphasize the need for multi institutional validation to allow broader applicability and resilience of the models to academically diverse environments.

In the future, research efforts to expand the robustness and the scope of automated CO–PO mapping systems should be made towards multilingual transformer language models, for example mBERT or XLM-R, to be applied to institutions that are not operating in the English context. Collaborative training in the universities, while preserving data privacy, is a promising path forward that is provided by federated learning. Furthermore, adaptive feature engineering, including semantic role labeling and clause level embeddings, can substantially increase model accuracy for statements describing multiple (complex or compound) outcomes.

To advance the field of AI in Education (AIEd), we urge the academic and developer communities to establish open benchmarks, standardized datasets, and reproducible evaluation frameworks for CO–PO mapping. By pooling resources—including codebases, annotated corpora, and alignment metrics—we can collectively accelerate the development of equitable, scalable, and high-impact solutions for automated curriculum alignment in outcome-based education systems worldwide.

References

ABET. (2023). *Criteria for accrediting engineering programs, 2023–2024*. ABET. https://www.abet.org

Agrawal, E., Tungikar, V., & Joshi, Y. (2021). Method for assessment and attainment of course and programme outcomes for tier-I institutes in India. *Journal of Engineering Education Transformations, 34*(3), 35–41.

Akramy, S. A. (2021). Implementation of outcome-based education (obe) in afghan universities: Lecturers' voices. *DergiPark (Istanbul University), 1*(6).

Amiruddin, M., Damit, A., Omar, K., Hazwan, M., & Puad, M. (2021). Issues and challenges of outcome-based education (OBE) implementation among Malaysian vocational college teachers. *International Journal of Academic Research in Business and Social Sciences, 11*(3), 2222–6990. https://doi.org/10.6007/IJARBSS/v11-i3/8624

Arun Kumar, B. R. (2021). Value-added courses with industry endorsement for bridging curriculum gap in the outcome-based learning: An effective strategy to post-graduate studies. *Journal of Education, 6*(6), 002205742097205. https://doi.org/10.1177/0022057420972056

Anderson, L. W., & Krathwohl, D. R. (Eds.). (2001). *A taxonomy for learning, teaching, and assessing: A revision of bloom's taxonomy of educational objectives*. Longman.

Biggs, J. (1996). Enhancing teaching through constructive alignment. *Higher Education, 32*(3), 347–364. https://doi.org/10.1007/BF00138871

Biggs, J., & Tang, C. (2011). *Teaching for quality learning at university* (4th ed.). McGraw-Hill/ Open University Press.

Da Cruz Alves, N., Gresse Von Wangenheim, C., & Martins-Pacheco, L. H. (2021). Assessing product creativity in computing education: A systematic mapping study. *Informatics in Education, 1*(3), 19–45. https://doi.org/10.15388/infedu.2021.02

De andrade, T. L., Rigo, S. J., & Barbosa, J. L. V. (2021). Active methodology, educational data mining and learning analytics: A systematic mapping study. *Informatics in Education, 3*(5). https://doi.org/10.15388/infedu.2021.09

Frenk, J., Chen, L., Bhutta, Z. A., Cohen, J., Crisp, N., Evans, T., Fineberg, H., Garcia, P., Ke, Y., Kelley, P., Kistnasamy, B., Meleis, A., Naylor, D., Pablos-Mendez, A., Reddy, S., Scrimshaw, S., Sepulveda, J., Serwadda, D., & Zurayk, H. (2020). Health professionals for a new century: Transforming education to strengthen health Systems in an Interdependent World. *The Lancet, 376*(9756), 1923–1958.

Gao, W., Liu, Q., Huang, Z., Yin, Y., Bi, H., Wang, M.-C., Ma, J., Wang, S., & Su, Y. (2021). RCD: Relation map driven cognitive diagnosis for intelligent education systems. *Proceedings of the 45th International ACM SIGIR Conference on Research and Development in Information Retrieval, 8*(5), 501–510. https://doi.org/10.1145/3404835.3462932

He, J. (2017). The historical evolvement of computer-assisted language teaching and its positive and negative effects. *DEStech Transactions on Social Science, Education and Human Science, 1*. https://doi.org/10.12783/dtssehs/ssme2017/13032

Kumar, K. S. A., Worku, B., Hababa, S. M., R, B., & A Y, P. (2021). Outcome-based education: A case study on course outcomes, program outcomes and attainment for big data analytics course. *Journal of Engineering Education Transformations, 35*(2), 63–72. https://doi.org/10.16920/jeet/2021/v35i2/22072

Lavanya, C., & Murthy, J. N. (2022). Assessment and attainment of course outcomes and program outcomes. *Journal of Engineering Education Transformations, 35*(4). https://doi.org/10.16920/jeet/2022/v35i4/22109

Marasigan, J. C. (2020). Validating and utilizing an outcomes-based education (OBE) student teaching manual in a State University in The Philippines. *Indian Journal of Science and Technology, 13*(07), 743–755. https://doi.org/10.17485/ijst/2020/v13i07/148722

NAAC. (2022). *Manual for accreditation of universities*. National Assessment and Accreditation Council. https://www.naac.gov.in

Owens, D. C., Sadler, T. D., Barlow, A. T., & Smith-Walters, C. (2020). Student motivation from and resistance to active learning rooted in essential science practices. *Research in Science Education, 50*(1), 253–277. https://doi.org/10.1007/s11165-017-9688-1

Raj, R., Sabin, M., Impagliazzo, J., Bowers, D., Daniels, M., Hermans, F., Kiesler, N., Kumar, A. N., MacKellar, B., McCauley, R., Nabi, S. W., & Oudshoorn, M. (2021). Professional Competencies in Computing Education. *Proceedings of the 2021 Working Group Reports on Innovation and Technology in Computer Science Education*, *6*(8). https://doi.org/10.1145/3502870.3506570.

Reddy, B. R., Natarajan Karuppiah, Asif, M., & S. Ravivarman. (2021). A case study on the assessment of program quality through CO-PO mapping and its attainment. Journal of Engineering Education/Journal of Engineering Education Transformations/Journal of Engineering Education Transformation, 34(0), 104–104. https://doi.org/10.16920/jeet/2021/v34i0/157114.

Shah, M., & Kolhekar, M. (2021). A case-study on leveraging the policies on outcome-based education. *Journal of Engineering Education Transformations, 35*(2), 126–139. https://doi.org/10.16920/jeet/2021/v35i2/22080

Spady, W. G. (1994). *Outcome-based education: Critical issues and answers*. American Association of School Administrators.

Sunitha, G., & Avanija, J. (2021). Program assessment and evaluation for continuous quality improvement. *Journal of Engineering Education Transformations, 34*(3), 42. https://doi.org/10.16920/jeet/2021/v34i3/150016

Trinidad, J. E. (2024). Positionality in the politics of education: Mapping the strategies and constraints of education reformers. *Educational Policy, 7*(3). https://doi.org/10.1177/08959048231220038

Wu, Y., Xu, L., & Philbin, S. P. (2023). Evaluating the role of the communication skills of engineering students on employability according to the outcome-based education (OBE) theory. *Sustainability, 15*(12), 9711. https://doi.org/10.3390/su15129711

Yang, P. (2020). Humanities education reform exploration and practice under outcomes-based education (OBE). *The Education and Science Journal, 22*(2), 78–97. https://doi.org/10.17853/1994-5639-2020-2-78-97

Zamir, M. Z., Abid, M. I., Fazal, M. R., Qazi, M. A. A. R., & Kamran, M. (2022). Switching to outcome-based education (OBE) system, a paradigm shift in engineering education. *IEEE Transactions on Education, 1*(2), 1–8. https://doi.org/10.1109/TE.2022.3169184

Advancing Outcome-Based Education Through Machine Learning: The Road Ahead

Subhra Prosun Paul, Shafikul Islam, and Shah Murtaza Rashid AI Masud

Abstract Outcome-based education (OBE) focuses on defining specific learning outcomes for students and continuously modifying teaching and assessment methods to achieve these outcomes. As one of the most transformative influences on education, the digitalization of the learning process is growing exponentially, and with it, the possibilities for new meaning through the application of Machine Learning (ML) for data-driven decision-making, adaptive and predictive learning can help to improve the efficiency of OBE. This chapter discusses the integration of ML and OBE as a collaborative agent for advanced algorithms that automate assessments, identify at-risk learners, provide real-time feedback, and personalize learning pathways. It also raises fundamental questions about data privacy, algorithmic fairness, and institutional willingness to adopt a new technology. The systematic roadmap for implementing ML-powered OBE frameworks includes strategic contingent plans, ranging from readiness assessment to full-scale deployment. The chapter concludes with a vision of a learning future characterized by intelligent, self-evaluating, student-centric, and lifelong education systems that yield measurable gains in the educational process.

Keywords Adaptive learning · Educational analytics · Machine learning · Outcome-based education · Predictive modeling

S. P. Paul (✉) · S. M. R. AI. Masud
Department of CSE, University of Asia Pacific (UAP), Dhaka, Bangladesh
e-mail: murtaza@uap-bd.edu

S. Islam
Department of CSE, Uttara University, Dhaka, Bangladesh
e-mail: shafikul.islam@uttarauniversity.edu.bd

D. Vetrithangam et al. (eds.), *Transforming Outcome-Based Education with Machine Learning*, https://doi.org/10.1007/978-981-95-4925-2_11

1 Introduction

Globalization, advances in technology, and the growing demand for employability and lifelong learning are key factors driving education systems worldwide to adapt rapidly to the evolving needs of the twenty first century. In today's fast-changing world, traditional education models that rely on teacher-centered instruction, rigid syllabi, and standardized assessments are no longer sufficient to meet the diverse needs of learners and the demands of modern industries. One of the most significant transformations in twenty first-century higher education is the shift to Outcome-Based Education (OBE). Instead of focusing on inputs such as syllabus coverage and classroom hours, OBE emphasizes clearly defined learning outcomes that students are expected to achieve by the end of a course or program. Typically, such outcomes align with professional competencies, industry needs, and national qualification frameworks (Aljohani, 2017). The core principles of Outcome-Based Education (OBE) include a clear focus on learning outcomes, designing curriculum and instruction based on those outcomes, setting high expectations for all students, and providing multiple opportunities for students to demonstrate what they have learned. Although the OBE, by its nature, offers benefits such as maximizing accountability, improving congruency in the academic curriculum, and promoting continuous improvement as a system of education, its adoption is not without its challenges. Some of these pain points include the manual process of mapping course outcomes (COs) to program outcomes (POs), which consumes a significant amount of time, a lack of access to actionable and real-time learning data, and a one-size-fits-all approach to instruction and assessment, among others. With the practical implementation of OBE within continuously innovating institutions, OBE's future-ready systems will be the need of the hour—intelligent systems that tame complexity and guide decision-making with data-driven insights and personalized learning experiences. This is where Machine Learning (ML), a strong branch of Artificial Intelligence (AI), comes into play in the education field (Batool, 2023). Machine Learning algorithms can analyze large-scale educational data to recognize patterns, make predictions, and facilitate personalized interventions. ML can assist you in identifying academic underachievers, predicting student performance, clustering and segmenting learners according to engagement or behavior, and even recommending customized learning materials.

The interplay between ML and OBE presents a due course for precision education, wherein, through dynamic feedback machines, the learning process is tailored according to an analytics-driven development for every learner on a pathway to a specific outcome. This becomes even more crucial as the trends in teaching (education technology) and learning (outcome-based education) converge to make teaching more efficient and assess a broader range of outcomes (outcome-based education) in a more transparent, measurable, and equitable manner. In this chapter, we examine current practices, explore where ML can enter and add value, and build a framework for scalable, ethical, effective implementation of ML in education through the lens of Outcome-Based Education (Feldstein & Hill, 2017). It also consults real-life case studies, practical challenges, and future directions, which may provide educators, administrators, and policymakers with the knowledge to use ML for improving OBE structures in higher education and beyond.

2 Literature Review

Outcome-Based Education (OBE) is a growing educational paradigm that focuses on ensuring students achieve key learning outcomes during their academic experiences. OBE emphasizes the desired, measurable outcomes for student programs rather than the instructional inputs (Davis et al., 2007) to improve how assessments relate to instructional practices. However, OBE is not without its challenges, including difficulties in tracking systems, limitations in real-time data analytics, and the challenge of personalizing education to meet the individual needs of students (Finkelstein & McCaffrey, 2019). Here, machine learning (ML), as a subfield of artificial intelligence (AI), can serve as a comprehensive solution to enable stakeholders to make data-driven decisions, customize learning experiences, and provide timely feedback to educators and learners (Baker & Yacef, 2009).

However, these studies primarily focused on predicting students' performance, recognizing learners' styles, and constructing adaptive learning platforms that applied supervised learning, unsupervised learning, and reinforcement learning, respectively (Chen & Zhang, 2020). Typically, in predictive analytics, supervised learning methods are used to model students' future performance based on their previous behavior (e.g., their performance in prior courses or their engagement) (Huang & Spector, 2021). In contrast, unsupervised learning approaches identify latent structures that relate student activities to personalize instruction, thereby improving educational experiences (Baker & Yacef, 2009). Using informative mechanisms for adaptation with verified technology for students, like reinforcement learning that monitors their performance and records their process on immediate stream(s), brings a solution for making learners' learning journey more sustainable and relevant, bringing modifications in content as per the in-depth progress of the learner itself (Deeva & Muravska, 2019). ML techniques can better inform OBE by generating insights based on student performance, engagement, and degrees of mastery of program and course outcomes.

OBE systems have already been tried out in some educational institutions with positive results regarding the incorporation of ML into them. For example, predictive analytics has helped identify students at risk for failure earlier in their school careers, allowing for interventions to be made sooner (Goe & Croft, 2016). For example, a case study by (Koh & Tan, 2018) reported the ability of various ML models to predict student performance and provide educators with the opportunity to optimize learning pathways and improve educational performance. Intelligent tutoring systems based on artificial intelligence (AI) and machine learning (ML) are also designed to provide tailor-made learning experiences tailored to the learning journeys of individual learners, thus conforming to the OBE philosophy (Müller &

Rößling, 2018). These systems utilize real-time data to deliver immediate feedback, enabling students to achieve their learning goals more efficiently.

Several ethical and practical issues need to be addressed, while the flexible linking of OBE with ML ensures promising results. One of the major concerns is data privacy, as educational institutions must adhere to regulations such as the General Data Protection Regulation (GDPR) and the Family Educational Rights and Privacy Act (FERPA) to protect student data (Popenici & Kerr, 2017). In addition, of course, there is a worry that algorithmic bias may be a serious problem; if ML models are trained on biased data, they can, by design, amplify existing inequalities (Finkelstein & McCaffrey, 2019). The following ethical challenges underscore the need for a robust ethical framework to guide the development and deployment of ML-based OBE systems. Hence, systems that are transparent, accountable, and fair will be essential for increasing the responsible use of ML in education (Nguyen & Chan, 2021).

3 Overview of Outcome-Based Education (OBE)

Outcome-Based Education (OBE) has become one of the most modern education methods, aiming to reshape the educational process by restructuring outputs, specifically learning outcomes that students are expected to know, do, and demonstrate, rather than focusing on inputs such as hourly teaching or syllabus completion. It is a goal-driven paradigm that creates alignment between the design of curriculum materials, the nature of instructional strategies, and the forms of assessment that share clear goals that students should consider. Figure 1 illustrates the steps for implementing OBE.

3.1 Definition and Philosophy

OBE is fundamentally an educational methodology where all elements of the teaching-learning process are oriented towards achieving the desired learning outcomes. These learning outcomes describe what students should be able to do after completing a course or program. The key principles of the Outcome-Based Education (OBE) philosophy guide the implementation of this approach in academic institutions. First, Clarity of Focus ensures that every instructional decision is consistently aligned with the specific outcomes students are expected to master, thereby establishing a clear vision for learning. Second, Backwards Design very deliberately considers how the curriculum guides teaching strategies toward long-term goals (which may extend beyond a year of study) and formative and summative assessments. Third, the High Expectations for All Learners philosophy reflects the belief that all learners can reach incredible heights if they engage in the appropriate level of learning rigor and if their differences are nurtured rather than ignored.

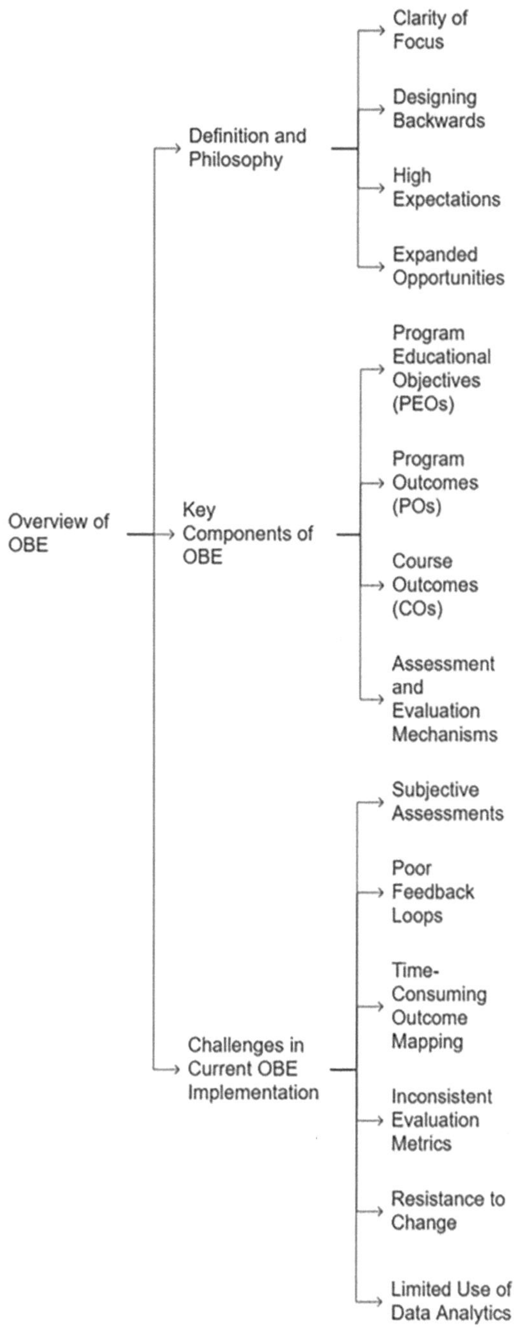

Fig. 1 Implementation steps of OBE

This guarantees fairness and inclusiveness in education and treatment. For that reason, they foster a shift away from rote learning to the development of twenty first-century skills—critical thinking, communication skills, collaboration, and problem-solving—that will enable learners to meet real-world challenges.

3.2 Key Components of OBE

Outcome-Based Education (OBE) is an educational approach that emphasizes measurable outcomes through a structured evaluation process and a hierarchical framework of outcomes. It ensures that students achieve specific competencies and skills aligned with program goals and societal needs. The key components of OBE include:

1) Program Educational Objectives (PEOs)
   PEOs are broad, long-term statements that describe the career and professional achievements graduates are expected to attain 3–5 years after completing the program. These objectives typically reflect professional growth, lifelong learning, and contributions to society and the profession.
2) Program Outcomes (POs)
   Program Outcomes (POs) are specific, measurable statements that describe the knowledge, skills, and attitudes students are expected to attain by the time they graduate from an academic program. These outcomes align with the expectations of national and international accreditation bodies, such as the National Board of Accreditation (NBA) in India, ABET in the USA, and the Washington Accord. POs act as critical indicators for achieving the broader Program Educational Objectives (PEOs) and are essential in evaluating the effectiveness of the curriculum. Typical Program Outcomes include the application of engineering knowledge, the ability to analyze problems, design and develop solutions, conduct investigations, and use modern tools effectively. Additionally, they encompass ethical responsibility, effective communication, project management skills, and a commitment to lifelong learning.
3) Course Outcomes (COs)
   COs define the specific skills and competencies that students are expected to acquire upon successful completion of a particular course. Each CO contributes to one or more POs and is carefully designed to align with the overall program objectives.
4) Assessment and Evaluation
   Assessment and Evaluation play a crucial role in ensuring the effectiveness of Outcome-Based Education (OBE). Both direct and indirect assessment methods are used to measure the attainment of course and program outcomes. Direct assessments involve measurable academic activities such as quizzes, assignments, laboratory work, mini-projects, and examinations. These are carefully mapped to various cognitive levels defined in Bloom's Taxonomy, including

Remember, Understand, Apply, Analyze, Evaluate, and Create. On the other hand, indirect assessments gather qualitative insights through feedback mechanisms such as student surveys, alum feedback, and peer reviews. Together, these assessment strategies provide a comprehensive understanding of student performance and learning outcome achievement, enabling continuous improvement in teaching and curriculum design.

3.3 Challenges in Current OBE Implementation

Although OBE offers a structured, outcome-focused approach, its implementation has several drawbacks. Outcome-Based Education (OBE) is often misused for various reasons, including issues such as a lack of guidance, an inappropriate framework, and academic rigidity. First, subjective assessment is an issue—admission personnel must often review applications individually, which can lead to unfair or inconsistent decisions (Feldstein & Hill, 2017). Second, feedback from students to teachers is usually slow or ignored, making it impossible to take rapid action to improve learning. Third, mapping outcomes—connecting course goals to program and institutional goals—is tedious and labor-intensive when done manually, and prone to error. Fourth, there are no common standards—the grading systems and achievement levels vary, and the results become confusing and unclear when coming from different departments. Fifth, resistance to change is another challenge—some teachers and staff want to continue as before and find OBE's data tracking and reporting to be too complicated or complex to learn. Finally, there is limited use of data. While OBE systems generate large amounts of data, many institutions lack the necessary tools or skills to analyze it thoroughly, making it difficult to improve teaching and learning. These issues underscore the need for more intelligent, automated systems, such as those based on machine learning, to elevate OBE to the next level.

4 Machine Learning in Education: A Primer

The sector around formal education delivery is also undergoing digital transformation, making it easier to retrieve knowledge, understand learning better, assess learning outcomes, and utilize associations to enhance the learning experience. Central to this change is Machine Learning (ML)—a subset of Artificial Intelligence that enables systems to learn from data and improve their performance over time, without the need for explicit task-specific programming. For education, ML paves the way for personalized learning, predictive analysis of students' performance, automated assessments, and data-driven institutional decisions, among other benefits. (Huang & Fang, 2013). Figure 2 further elaborates on ML strategies used in educational guidance.

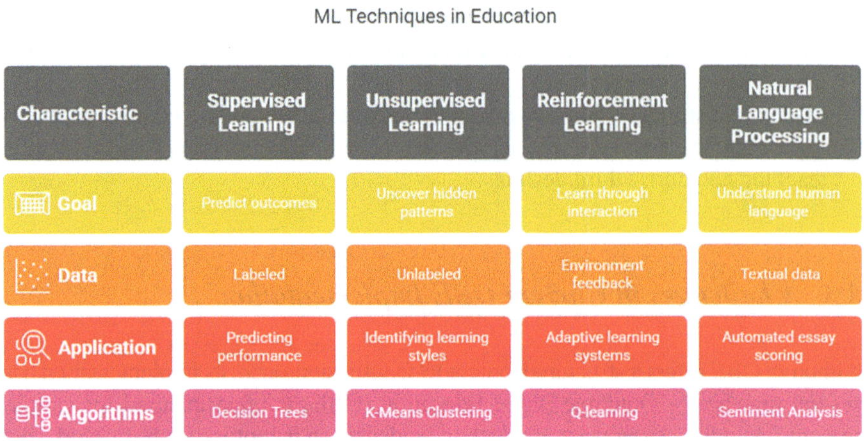

Fig. 2 ML techniques in education

Machine learning involves applying skills and capabilities to develop models that undergo continuous improvements. Instead of having static instructions, ML models identify complex relationships in data, create mathematical models of them, and then utilize these models to make predictions or informed decisions. The process of teaching and learning using Machine Learning (ML) in education follows these steps. First, data will be collected from various sources, including students' scores, attendance records, online activities, and discussion forum posts. Then, the features, or components of the data, that are most relevant for identifying or predicting student performance are selected. Then, a model is trained on past data and algorithms to learn patterns and behaviors. Steps include, but are not limited to, cleaning up the training dataset with minimal input after receiving it, using the training data to train a model that will be used for predictions or insights, and validating the model with test data. ML then uses such data to interpret and analyze the actions, engagement levels, performance, and learning styles of students. Consequently, it helps teachers and schools develop more effective and tailored teaching approaches.

4.1 Common ML Techniques in Education

Machine Learning methods can be broadly classified into different categories based on how they learn from data. Each technique serves different educational functions, and its effectiveness depends on the context and quality of available data.

4.1.1 Supervised Learning

Whenever you have data with a corresponding label, it can be used to train a model using supervised learning. In education, for example, a supervised learning algorithm is used to predict student performance by providing a model with past data, such as exam grades, assignment performance, and attendance (Nguyen & Chan, 2021). This data is then used to train the model, which learns the patterns and can, therefore, predict either a student's final grade or their likelihood of passing a course. It also fuels early warning systems, where the model identifies students most at risk of failing and generates alerts so that teachers can step in before it is too late. Its primary function, however, is to provide personalized recommendations, suggesting learning materials based on a student's prior performance and anticipated challenges. Popular Algorithms: Decision Trees, Support Vector Machine (SVM), Random Forest, Logistic Regression & Neural Networks. Suppose the model trained on years of historical academic data indicates that Student X is likely to fail a math test with a 75% likelihood. In that case, the system can instantly alert the teacher, allowing them to intercept the failure before it's too late.

4.1.2 Unsupervised Learning

It works with unlabeled data, which means that in this type of machine learning model, it learns patterns and structures independently without the introduction of correct output. In education, this technique can provide insights into how students interact with learning materials—such as the number of words they read, the number of people who attempt quizzes, or the level of activity in discussion forums. The model can group, or cluster, students based on their behaviour or learning styles. For instance, it could identify patterns that reveal some students prefer visual content, while others prefer audio materials or hands-on activities. Grouping similar courses or modules with the same learning outcomes helps to improve course design as part of topic optimization. Some standard algorithms for unsupervised learning include k-means clustering, hierarchical clustering, and principal component analysis (PCA). For instance, if we use student interaction data to classify students into different learning style groups (visual, auditory, or kinesthetic learners), we can recommend individual learning resources that are well-suited for each group.

4.1.3 Reinforcement Learning

Reinforcement Learning (RL) is a type of Machine Learning in which a model interacts with its environment and learns through the experience of receiving a reward or a penalty for its actions. In education, RL can be applied to build systems that adaptively teach students lessons tailored specifically to them. For example, based on a student's performance, the system can select the next activity or lesson, adjusting in real-time. Students earn points/rewards and progress through levels or learning modules, much like a game, thereby making learning more enjoyable by incorporating RL into game-like learning types. Popular RL algorithms include Q-learning, Deep Q Networks (DQN), and Policy Gradient Methods. Example: This is an e-learning platform where an RL agent provides students with increasingly complex tasks. When a student succeeds in a task, they move on to a more challenging one. If they have struggles, the system assigns more manageable or supportive tasks. This enables students to learn at their own pace, keeping them motivated.

4.1.4 Natural Language Processing (NLP) in Education

Natural Language Processing (NLP) is an application of Artificial Intelligence (AI) that enables computers to comprehend and process human language (Romero & Ventura, 2010). This can be particularly useful in education, where NLP allows machines to read what students write and respond accordingly. One way NLP detectors are commonly used is in automated essay scoring, where an NLP model evaluates essays for aspects such as grammar, structure, vocabulary, and the student's ability to answer the question effectively. It can provide students with real-time feedback, guiding them in improving their responses to open-ended questions or in online discussions. NLP is also used in chatbots—brilliant virtual tutors that can answer student questions, offer hints, and recommend relevant materials (Baradari et al., 2025). Some widely used NLP techniques include sentiment analysis, Named Entity Recognition (NER), Transformers (such as BERT and GPT), and Text Classification. For instance, an NLP system can scan thousands of essays in a matter of seconds, identify those that lack critical thinking or clarity, and provide suggestions on how to improve.

4.2 The Growing Impact of ML in Education

With the introduction of digital learning platforms and Learning Management Systems (LMS), ML began to dominate the field of education. The Learning Pathways: Change the way learners learn by adapting to their performance, style, and pace with adaptive platforms. ML can track behavioral signals all the way to test-taking, allowing instructors to quantify levels of attention, motivation, or fatigue, and adapt strategies as needed (Improved Engagement Analytics). In Education, data insights can be an enabler of inclusion by identifying and addressing systemic issues, such as bias, inequities in resources, and disparities in academic achievement between groups (Siemens & Long, 2011). This Artificial Intelligence architecture may aid in medical diagnosis only if there are a variety of diverse and high-quality datasets, and then, if we have selected the correct models. Even after all this, if humans are involved in every step, we can avoid becoming too dependent on algorithms. Still, the algorithms are being used throughout the facility.

5 Synergies Between Machine Learning and Outcome-Based Education (OBE)

This union of ML and OBE is a significant value addition that enhances the accuracy, scalability, and effectiveness of today's education system. OBE emphasizes clear learning outcomes and ongoing assessment of student performance. In contrast, ML has the potential to analyze large volumes of educational data, uncovering patterns, forecasting trends, and automating critical processes. This will result in intelligent education ecosystems that are learner-centric, data-driven, and outcome-oriented. Table 1 below depicts the contribution of ML techniques to various aspects of OBE:

These synergies serve as a foundation for more adaptive and intelligent educational frameworks that ensure outcome alignment while continuously enhancing the learner experience.

5.1 *Predictive Analytics for Outcome Achievement*

Predictive analytics is one of the most valuable ways machine learning (ML) is helping in Outcome-Based Education (OBE). ML models can utilize past and real-time student data to predict the likelihood of a student achieving the expected Course Outcomes (COs) and Program Outcomes (POs). This data comes from various sources, including attendance records, quiz and assignment scores, time spent on learning platforms, participation in discussions or forums, and overall engagement and behavior. With these insights, institutions can take proactive steps to support students and enhance learning outcomes. The tool adapts the inferred information to set the likelihood of any student, using machine learning (ML) algorithms (logistic regression, random forests, deep learning models, etc.), predicting the chances of any student obtaining the outcome. For instance, a university might develop a predictive model using "Big Data" obtained from students' actions and activities on a virtual learning platform. Suppose it predicts a student is at risk of not meeting certain COs. In that case, it can raise a flag for teachers, who can then intervene with the individual student through tutoring, remedial modules, or in-person sessions (Slade & Prinsloo, 2013).

Table 1 Contribution of ML techniques to various aspects of OBE

OBE Element	ML Contribution
Mapping COs to POs	Automated classification and mapping models
Assessment & Evaluation	Predictive analytics, automated grading
Student engagement	Learning analytics, behavior & emotion detection
At-risk identification	Early warning systems using classification algorithms
Curriculum design	Data-driven course optimization and feedback loop

This not only drives the success of such programs at a high probability rate but also champions the essence of OBE in data-driven decision-making.

5.2 Intelligent Tutoring Systems (ITS) for Personalized Learning

AI tutors, also known as Intelligent Tutoring Systems (ITS), are systems driven by artificial intelligence that dynamically and individually adapt learning content and activities to the learner's pace, understanding, and preferences. By leveraging reinforcement learning, supervised learning, and natural language processing, adaptive learning systems contribute to the next stage in the evolution of pedagogical software. The systems are intelligent enough to monitor each student's progress against Course Outcomes (COs) and Program Outcomes (POs), shortlisting learning material suitable for the respective individual. When a student fails, the system can lower the content level, provide instant feedback, and suggest additional resources, such as videos, hints, or practice exercises, to complement the content. Unlike conventional teaching methods, which employ a "one-size-fits-all" approach, these systems ensure that each student receives the individualized help necessary to fully understand the topic. For instance, if a student struggles to grasp a programming concept during a computer science course, the system might automatically deliver simplified explanations, more frequent practice, and guiding tutorials. As a student progresses, the system adjusts the lessons accordingly to help them achieve the defined learning objectives. In this way, the adaptive learning technologies aid Outcome-Based Education (OBE) by enabling each student to master the requisite skills, thereby enhancing the overall performance of the program.

5.3 Sentiment and Feedback Analysis Through NLP

One thing that is easy to overlook, yet crucial for OBE, is the need to collect and analyze feedback from students on their learning experience, so that the learning experience can continually evolve. Traditional approaches, such as end-of-semester surveys, often suffer from limitations in scope, frequency, and depth of coverage. The conventional approach has been rendered ineffective due to its limitations in scaling. By using ML models, we can decode: (1) Sentiment Analysis, (2) Text Classification, (3) Keyword Extraction. Summary of open-ended student feedback. Students can comment in online discussion forums, complete course review forms, or post comments on social media. Such feedback is beneficial for enhancing courses. For example, it can help identify common problems or compliments, perceive the emotional tone of the students, and understand how students feel about the effectiveness of the course. It also helps assess whether teaching is achieving the desired learning outcomes. After one semester, NLP algorithms can parse tens of thousands of student comments and cluster them into themes such as "fairness of assessments," "clarity of teaching," or "workload." This helps curriculum designers and academic committees overcome shortcomings and ensure that the programme's learning objectives are being fulfilled. Additionally, it can detect when students are feeling frustrated or upset, providing instructors or counselors with an opportunity to reach out and offer support, to improve student retention and success.

5.4 Enhanced Outcome Mapping and Rubric Standardization

One key assumption of OBE is that COs are systematically mapped to POs, which can be a tedious and error-prone task when performed manually. This process can be automated with ML, which would, based on past data, course structure, and course content, provide suggestions on mappings. Similar outcomes across courses can be grouped using clustering techniques (Sukhnandan & Underwood, 2019). Learning verbs and knowledge domains (in alignment with Bloom's Taxonomy) categorization (data): text classification models can classify and organize possible outcomes. Use association rule mining to find the relationship between COs and POs of students to find the pattern in student performance data. For instance, any institution may utilize ML models to analyze numerous historical CO-PO mappings, along with student performance data. This will facilitate system-driven recommendations for the most relevant PO for each newly created CO, thereby ensuring consistency and minimizing manual work. ML can also help with creating standardized grading rubrics by learning from past grading styles. This minimizes evaluator bias and ensures more fair and uniform evaluations of courses.

5.5 Curriculum Design and Optimization

Curriculum design is a strategically nuanced and often delicate process of balancing learning objectives, the complexity of the course material, sequencing, and student feedback. ML can render the process more evidence-based by: Courses with extremely low outcome attainment. Units during which students disconnect or leave. Coverage of topics missed or then covered in the curriculum. For example, a neural network that learns from historical student data can then emulate how curriculum changes (e.g., changing the order of courses or adjusting credit weights) impact overall program outcome attainment (Tondeur & van Braak, 2017). This will help the curriculum designers make evidence-based changes consistent with OBE objectives.

5.6 Real-Time Student Engagement Monitoring

Behavioral analytics, emotion recognition (through webcam or sensors), and interaction logs can all power Machine Learning models that identify real-time engagement. This contributes to the Real-Time adaptation of teaching methods. Instant notifications to instructors about unengaged students. Tailored nudges or motivational content to improve participation. Aligning engagement analytics with how well students are tracking their learning outcomes keeps the teaching process responsive and effective.

6 Case Studies and Applications

Through various case studies, we can see how machine learning is already being utilized to enhance student performance, identify students at risk of dropping out, and facilitate timely and effective interventions. It also enables better and faster decision-making, particularly in the context vof Outcome-Based Education (OBE).

6.1 Case Study 1: Identifying at-Risk Students Early

Real-Time Problem: A private engineering institution was facing a persistent challenge in identifying students who were at risk of academic failure or potential dropout. The root of the problem lay in the inability to detect early signs of declining academic performance and low engagement levels. As a result, many students were falling through the cracks without receiving the timely support and intervention they needed to succeed. This gap highlighted the urgent need for a data-driven approach to proactively monitor and support at-risk students, ensuring better academic outcomes.

Outcomes and retention.

Objective The goal was to develop a machine learning-based early warning system that alerts educators to students in a portfolio based on multiple academic and behavioral data points.

Implementation The college utilized various data sources, including attendance records, assignment scores, quiz marks, LMS login frequency, and forum participation. After cleaning and transforming the data, a Random Forest classification model can be used to classify the students into "at-risk" and "not at-risk" groups. The model highlighted LMS activity and mid-semester test scores as the leading indicators of student risk. Reports generated by the model were then provided to faculty members to enable early intervention through mentoring and support.

Outcome By identifying students with K-12 grades who indicated they would drop out of higher education, this machine learning system enabled faculty members to target counseling and academic support resources for students who most needed them. This resulted in an 18% reduction in the failure rate over two academic sessions. Specifically, the system played a direct role in achieving better course outcomes (COs) as well as overall student retention, both of which are clear goals in OBE.

6.2 Case Study 2: Automated Assessment and Feedback System

Real-Time Problem: A significant challenge faced by an academic institution was the manual grading of student essays and assignments, which proved to be time-consuming, inconsistent, and inefficient. Instructors struggled to deliver timely, personalized feedback due to the high workload and limited resources. This delay in feedback negatively impacted the quality of learning and slowed down the students' ability to improve their performance effectively. The institution recognized the need for an automated solution that could streamline assessment processes while maintaining accuracy and providing meaningful, individualized feedback.

Objective: Your goal was to develop a machine learning-based system that could automatically grade student submissions and provide helpful comments, thereby alleviating the load on teachers and ensuring fair grading.

Implementation: Note that it was trained on a dataset of responses, which it scored based on fluency, grammar, the relevance of the response to the given prompt, and critical thinking. Using Natural Language Processing (NLP) techniques, the model leveraged this information to generate automatic scores and personalized recommendations for improvement. It also tracked revisions through several drafts, prompting students to rework and refine their writing.

Outcome: "Automated grading cut assessment time by 70%, freeing teachers to spend more time on instructing." It increased transparency and reduced variability in grading, thereby promoting progressive improvement in student writing. Solution leveraged end-to-end continuous evaluation based on COs, eliminating human bias and supporting OBE-compliant standards. These examples provide a glimpse into how machine learning is transforming education today. ML is enabling timely intervention and supporting the automation of assessments, which are helping institutions worldwide achieve better outcomes, sustain academic quality, and personalize learning. It verifies that ML does not need to look like something beyond tomorrow; it is a practical tool we have today.

7 Challenges and Ethical Considerations

Although the application of machine learning in outcome-based education (OBE) holds great promise, it is also surrounded by several challenges and ethical issues that hinder its adoption (Zhu & Shen, 2021). Without this being done, ML applications in education will remain inequitable, ineffective, and non-compliant to privacy standards. Subsequent sections discuss major issues, such as data privacy, algorithmic bias, infrastructure readiness, and faculty preparedness.

7.1 *Data Privacy*

Handling Utilities of Sensitive Data. One of the biggest challenges of ML deployment in educational institutions is the type of data it handles. Institutions must comply with rigorous data privacy standards to protect learners' personal and academic information. In education, machine learning is based on thousands of student data points, so you must also follow stringent privacy guidelines. Note that two critical frameworks are the General Data Protection Regulation (GDPR), which governs the European Union, and the Family Educational Rights and Privacy Act (FERPA), which governs the United States. GDPR mandates that organizations clearly explain how they use personal data, obtain students' consent before collecting it, and provide the ability to view, delete, or decline the use of their data. FERPA, the law that protects students' educational records, restricts access to this information without the student's consent. A serious concern is that ML-based systems can collect vast amounts of data (e.g., attendance, grades, and emotional responses), which can become very risky in the event of a data breach. Additionally, many schools share this data with third-party platforms, thereby increasing the opportunities for leaks. To address these issues, institutions must encrypt data both during transit and at rest. Additionally, they should apply privacy-by-design principles when developing ML systems and conduct periodic audits and checks to ensure compliance with relevant privacy laws. These measures ensure that student information is protected while still enabling educators to leverage ML tools.

7.2 *Algorithmic Bias*

An Algorithmic bias is one of the grave concerns while implementing ML in education. ML models are typically trained on past data, which can introduce bias in education, including disparities in access, implicit bias among instructors, or systemic inequities within the education system. If undetected and unaddressed, these biases can result in ML models that entrench or even exacerbate existing inequalities (Idowu et al., 2024). *Regular auditing of machine learning models used in education can help ensure fairness and the absence of biases. These checks, referred to as bias audits, help determine whether the model is making biased decisions that might negatively impact specific groups. The data used to train the model may not accurately represent the entire population, but rather only reflect a particular subset of it. This means that the dataset must be a mixture of students' backgrounds, allowing your model to learn accordingly and behave reasonably for all. An additional critical component is the explainability—the understanding of how and why a machine learning model arrives at its decisions. The first is that when a model is more transparent, it is easier to identify and correct biased behavior. Using techniques such as Explainable AI (XAI), the model aims to make machine learning decisions more transparent and understandable to institutions, ensuring they can be assured that the technology is making the right and justified choices.*

7.3 *Infrastructure and Faculty Preparedness*

ML has enormous potential, but there are very few educational institutions that have proper infrastructure and good faculty available to teach it. Technical infrastructure is one of the most significant challenges for the application of machine learning (ML) systems in educational institutions. This is often the case, as various organizations—especially in less developed regions—lack the computing power, internet connectivity, or data repositories required for the deployment and functioning of ML models. Another problem is that the high cost of adopting ML is a significant obstacle, in addition to infrastructure shortcomings. The costs associated with acquiring advanced software, PCR, and powerful servers, as well as hiring skilled data scientists, can be prohibitive for many lower-resourced institutions. Moreover, implementing ML tools alongside existing systems, such as the Learning Management System (LMS) and student information systems, is complex and requires a high degree of customization. This internal decision-making process can be challenging, as it involves striking a balance between the need for technical expertise and the goal of creating a cohesive and efficient digital learning ecosystem.

7.3.1 Faculty Readiness

Most teachers lack sufficient knowledge and technical skills required to utilize machine learning (ML) tools in their teaching. As a result, they may not use these tools to their fullest potential, and a gap can arise between the tools' capabilities and their actual application. Some educators may also be nervous about ML because they are less sure it will respect their agency or alter their pedagogical style too significantly. Likewise, there are few training opportunities available on how to utilize ML in education effectively. However, with minimal re-imbursement for upskilling in what these new tools and new pedagogies entail, keeping up is challenging for faculty.

Such challenges would eventually require schools, especially universities, to invest in better technical infrastructure. Remember, cloud services enable you to keep costs low while maintaining sufficient computing power at your disposal. Partnering with EdTech can help gain new ML tools and provide valuable technical assistance. First, universities/work academies need to create training programs for teachers so that they know how to deploy a responsible ML-based solution. This might involve offering workshops, online courses, webinars, and peer-to-peer learning groups to promote ongoing professional development.

7.3.2 Ethical Implications of Surveillance and Autonomy

Some systems use machine learning techniques, such as emotion recognition and engagement tracking, to monitor disengagement; although helpful, such systems raise ethical concerns about surveillance. The moral issues include the continuous monitoring of student activity, engagement, and emotional states, which can present ethical quandaries related to surveillance. Concerns exist over whether students should even consent to such monitoring and whether the data collected will be used responsibly.

Amidst these challenges lies a third factor: the shortage of personnel trained in machine learning (ML) and the faculty's reluctance to adopt even the most basic tools. Although powerful, most teachers are unsure about how to utilize ML tools. This means many of the utilities introduced are only partially or selectively adopted. Educators keen on maintaining stringent control in their classrooms may also be reluctant to adapt their teaching methods. To address these challenges, institutions must invest in enhanced infrastructure, explore cost-effective alternatives such as cloud-based systems, and establish robust partnerships with EdTech companies for access to tools and technical support. With that being said, ML-enabled institutions and departments should, above all else, also provide professional development opportunities (e.g., workshops, webinars, online courses) that teach faculty how to effectively and ethically integrate ML tools into their teaching. Table 2 highlights the Challenges and Ethical Considerations in ML-Enabled OBE.

However, the successful application of ML for OBE necessitates addressing ethical, privacy, and infrastructural issues. It requires a balanced approach to technology by institutions that can lead to better educational outcomes through the use of technology, while protecting fairness, equity, and students' rights.

8 Roadmap for ML-Enhanced OBE

It also requires a structured and strategic approach when integrating Machine Learning (ML) with the adoption of Outcome-based Education (OBE). A phased implementation roadmap can guide institutions in adopting ML-enhanced educational initiatives at a scalable and sustainable level. The key milestones for the

Table 2 Challenges and ethical considerations in ML-enabled OBE

Category	Challenges	Ethical Concerns	Proposed Solutions
1. Data privacy	– Large-scale data collection (e.g., LMS, biometric)– Data sharing with third-party vendors– Insufficient anonymization	– Violations of GDPR, FERPA– Lack of informed consent– Risk of data breaches	– Use of encryption & secure transmission– Privacy-by-design in ML tools– Regular audits & governance frameworks
2. Algorithmic bias	– Biased training data (gender, socioeconomic, cultural)– Lack of diverse representation– Inaccurate predictions for minority groups	– Discrimination & reduced opportunities– Reinforcement of systemic inequalities	– Regular bias audits using fairness metrics– Use diverse, representative datasets– Incorporate explainable AI (XAI) techniques
3. Infrastructure readiness	– Inadequate computing resources– High deployment costs– Difficult LMS/ML tool integration	– Unintentional exclusion of low-resource institutions	– Cloud-based ML solutions to reduce costs– Strategic investments in infrastructure– Collaborations with EdTech providers
4. Faculty readiness	– Lack of ML literacy among educators– Resistance to change– Insufficient training opportunities	– Underutilization of ML capabilities– Poor alignment with teaching goals	– Continuous faculty training programs– Incentivize ML adoption through support networks– Curriculum redesign to include AI pedagogy awareness
5. Surveillance & Autonomy	– Use of emotion detection, engagement tracking– Monitoring through biometric devices– Constant data capture during learning activities	– Intrusion into personal privacy– Reduced student autonomy and trust	– Informed consent with opt-out rights– Transparency about monitoring tools– Oversight via institutional ethics boards

roadmap encompass four essential phases: Readiness Assessment, Pilot Implementation, Full-Scale Integration, and Policy and Ethics Framework. Here, we provide a more detailed outline of each step, including key considerations and items to check at each stage.

8.1 Phase 1: Readiness Assessment

Data maturity, infrastructure, and faculty readiness are the three underlying prerequisites that an institution must address before implementing machine learning (ML)-driven Outcome-Based Education (OBE) systems. The first step is to review the institution's data environment—quality, completeness, and access to student data (considering the documentation of performance records, attendance, LMS engagement, and demographic factors). Then, institutions need to assess their current infrastructure (data storage systems, governance policies, and security protocols) and ensure it meets the basic needs for ML integration. A gap analysis can identify deficiencies in sustaining data to obtain the information required to deliver value. Training faculty and staff is also key." Introduce fundamental ML concepts and their usage in the education domain, along with ethical aspects. Faculty engaging with ML tools should employ strategies that facilitate their integration into the teaching and learning processes, such as interpreting the results to enhance teaching. You can also offer workshops to help them integrate ML-driven feedback into course design and teaching strategies. One primary objective is to develop an understanding of the institution's current capabilities and ensure that faculty and staff have the skills and mindset necessary for ML to thrive.

8.2 Phase 2: Pilot Implementation

The next step after the readiness assessment is a small pilot of ML in OBE. The objective is to test and refine models, resolve any challenges upfront, and obtain feedback from key stakeholders (Toxigon, 2025a). Here are some ideas in no specific order: Get Started with Small-Scale ML Projects. Identify a small pilot project that can be used to foster early program usage, like CO-PO mapping automation, automated assessment grading, or building a student performance predictive model. They also emphasize narrowing down use cases and demonstrating tangible improvement in how students perform, such as whether to use ML to study whether attendance and assignment scores can predict at-risk students. Track Results and Improve Processes. Pilot model success can be measured by comparing the expected student outcome with actual student success. Invite feedback from faculty, administrators, and students to reflect on strengths and areas for improvement. DSML (Data Science + another ML) Isolation to increase performance + limitations discovered in pilots (Toxigon, 2025b). Now, source Allocation and Support. Respond and provide dedicated technical support to troubleshoot any problems with the ML tools or infrastructure during the pilot. Regularly assess what resources will be needed for the pilot project in terms of data storage and retention, computing power, and time to train faculty and staff. There are several overarching aims of introducing machine learning (ML) tools in education, the key points of which involve implementing these tools in practice to understand the practical value of their, albeit somewhat ill-reputed, computational nature. Pilot ML in controlled environments to ensure institutions can track individual tools to explore their capabilities and assess their impact within teaching and learning. This will also help identify shiny objects, which are system challenges or constraints that should be considered and accounted for in long-term planning, allowing institutions to update their strategies and determine what needs to change to prepare for implementing ML solutions at scale. At this slower pace, ML tools can be effectively embedded in the teaching context, making for a smooth introduction.

8.3 *Level 3: Full-Scale Integration*

The steps towards formally embedding ML into the broader educational process within the institution can be initiated once the pilot phase has successfully evidenced the merits of ML for improving OBE. The deployment of machine learning models at scale and the establishment of feedback loops for continuous improvement are key aspects of this phase. The steps involve using predictive analytics and Adaptive Systems. Preemptively use predictive analytics to spot trends, e.g., students at risk of dropping out, underperforming, etc. Introduce adaptive learning platforms to ensure that educational content adapts according to individual student performance, allowing them to build on their tailored learning experience on the go towards their personal goals (Toxigon, 2025c). Use immediate feedback tools that provide faster instructor-to-student accountability and facilitate earlier intervention—scalable Data Processing. Greater use of ML may shift institutional demands for processing data. This may stem from the delivery of cloud-based services or enhancements to the infrastructure side of things. 'The question, therefore, is how to implement ML successfully within OBE systems, whilst the institutions must invest more in data privacy and security compliance with privacy and data protection laws. Thus, as ML systems advance to become part of everyday educational processes, they should enable reflection on the OBE frameworks already embedded in these processes. Performance and effectiveness need constant monitoring and tuning. Once ML scales for an institution, it may be required to invest in a more efficient data processing mechanism—either a cloud service or a more advanced on-premises infrastructure. As these systems advance, the protection of data and compliance with relevant laws should always remain a top priority. There are two main objectives: both strive to embed ML system performance into daily educational processes and to monitor and calibrate them during their lifetime to improve the added value towards learning score improvement.

8.4 *Phase 4: Support by Policy and Ethics Framework*

Position and Deploy Policies to Ensure Ethical Use of ML in Education: The final pathway in our roadmap is all about positioning and deploying policies to ensure ethical use of ML in education. It's an important step to achieving a balance between growth, new technologies, and the means of protecting the rights and interests of stakeholders. Key Action 1: Institutional AI Ethics Policy Draft: As one key action, implementing this enables the institutions to implement a policy that can be used to advise and guide ML models' development, deployment, and screening. It should account for data privacy and protection, transparency, explainability of algorithms, and a minimization of bias to ensure equity. You will not be able to develop this document in isolation—it should be done in consultation with a variety of stakeholders (legal counsel, an ethics review board, faculty, staff, and student

Table 3 Roadmap phases

Phase	Key Activities	Goals
Phase1:Readiness assessment	Evaluate data maturity, faculty training, and infrastructure	Prepare the institution for ML implementation
Phase 2:Pilot implementation	Start small-scale ML projects and refine models	Test ML tools in real-world contexts
Phase3:Full-scale integration	Deploy predictive analytics and adaptive learning systems	Integrate ML systems across the institution
Phase 4: Policy and ethics framework	Develop an AI ethics policy, ensure transparency and fairness	Ensure responsible and ethical use of ML

representatives) to articulate an approach that emphasizes transparency, accountability, and neutrality.

The next thing is setting up crystal-clear data governance procedures. This involves sharing transparency standards for model development, documenting training decisions, and more to guarantee that model outcomes are readable and comprehensible. The policy also needs to stress fostering accountability by making existing mechanisms accountable and equitable in collecting data, deploying models, and having feedback from model users. Ensuring that the ML model is not generating disparate and unforeseen results can only happen if we have a continuous monitoring process in place, which tweaks the ML model as per the requirements.

Another important aspect is compliance control. Each application should be verified by the ML application teams themselves that use an ML application that complies with national and international regulations (GDPR, FERPA, etc.) as well as institutional data governance policy. ML tools' ethical implications in educational settings require ongoing DEI (Diversity, Equity, and Inclusion) training for all faculty, administrators, and students. The fact is, we do (end up) with attached approval policies, which are approved to ensure the use of ML, but ethically in the context of the OBE setting. Roadmap Phases are detailed in Table 3.

This roadmap will help educational entities to navigate the different phases of ML implementation in a structured manner, encouraging a gradual and responsible adoption of ML so as to balance between innovation and responsible practice. A defensive strategy here would be to take a phased approach to this investment, as this helps to manage risk and allows gradual adoption while entities refine their learnings over time as they optimize their ML use cases.

9 Conclusion

Combining Outcome-Based Education (OBE) and Machine Learning (ML), the utilization of the latter can challenge the conventional paradigms of Measurement, Support & Achievement of learning outcomes in educational institutions. ML and OBE frameworks can be integrated to automate complex processes, rapidly identify students who are falling behind, delineate learning paths for them, and accumulate data on individual students that can inform teaching strategies. These capabilities are the little steps towards transforming education systems to be more inclusive, responsive, and impactful. However, this new transition is not without its own challenges. The potential of these tools is enormous, and their adoption needs to be made judiciously on issues of data privacy, algorithmic bias, building a standard data infrastructure, faculty training, etc. A structured, phased approach, integrating consideration of innovation with ethics and logistical feasibility, is required by institutions. As such, cross-disciplinary teams of educators, data scientists, and policymakers will be imperative for the successful implementation of ML-powered OBE systems.

The ML-OBE integration can be advanced further through future research and implementation efforts in the following areas. Federated Learning Models is a type of learning model that ensures that students' data information is kept safe and secure, as it will never leave the user's system, while allowing for collaborative intelligence to be shared across institutions. Use of Digital Twins in Education: Simulate individual learner profiles using digital twins to predict outcomes and to test interventions in a virtual environment before applying the same in the real one.

References

Aljohani, N. R. (2017). A comprehensive review of the major studies and theoretical models of student retention in higher education. *Higher Education Studies, 7*, 1–18.

Batool, S., Rashid, J., Nisar, M. W., Kim, J., Kwon, H. Y., & Hussain, A. (2023). Educational data mining to predict students' academic performance: A survey study. Education and Information Technologies, 28(1), 905–971.

Baker, R. S., & Yacef, K. (2009). The state of educational data mining in 2009: A review and future visions. Journal of Educational Data Mining, 1, 3–17.

Baradari, D., Kosmyna, N., Petrov, O., Kaplun, R. & Maes, P. (2025). NeuroChat: A neuroadaptive AI chatbot for customizing learning experiences. arXiv. https://arxiv.org/abs/2503.07599

Chen, J., & Zhang, Y. (2020). Machine learning in education: Applications and challenges. *International Journal of Educational Technology in Higher Education, 17*, 21–35. https://doi.org/10.1186/s41239-020-00217-0

Davis, M.H., Amin, Z., Grande, J.P., O'Neill, A.E., Pawlina, W., Viggiano, T.R. and Zuberi, R., 2007. Case studies in outcome-based education. *Medical Teacher, 29(7), 717–722.*

Deeva, T., & Muravska, T. (2019). Machine learning for student performance prediction in educational systems: A survey. *Procedia Computer Science, 162*, 348–355. https://doi.org/10.1016/j.procs.2019.11.075

Feldstein, M., & Hill, P. (2017). Learning with analytics: The emergence of a new learning ecosystem. *EDUCAUSE Rev., 52*, 40–57.

Finkelstein, J., & McCaffrey, R. (2019). Artificial intelligence in higher education: The challenges and opportunities for implementation. *International Journal of Artificial Intelligence in Education, 29*, 523–540. https://doi.org/10.1007/s40593-019-00184-1

Goe, L., & Croft, A. (2016). Building the teacher workforce through outcome-based education and data-driven assessment. *Journal of Teacher Education, 67*, 223–234. https://doi.org/10.3102/0034654316625988

Huang, R., & Spector, J. M. (2021). Exploring the role of predictive analytics and data mining in supporting educational decision-making. *Journal of Educational Computing Research, 59*, 607–625. https://doi.org/10.1177/07356331211005410

Huang, S., & Fang, N. (2013). Predicting student academic performance in an engineering dynamics course: A comparison of four types of predictive mathematical models. *Computers in Education, 61*, 133–145.

Idowu, J. A., Koshiyama, A. S., & Treleaven, P. (2024). Investigating algorithmic bias in student progress monitoring. *Computers and Education: Artificial Intelligence, 7*, 100267.

Koh, S. G., & Tan, A. L. (2018). Predicting student success with machine learning: A case study on data analytics for educational outcomes. *Educ. Data Min. J., 12*, 35–45.

Müller, M. & Rößling, G. (2018). Personalized learning in outcome-based education: Leveraging machine learning for adaptive learning systems. In *Springer Int. Publ.* https://doi.org/10.1007/978-3-319-89613-9_7

Nguyen, A., & Chan, A. (2021). Assessing the integration of artificial intelligence in education: Current trends and future directions. *Educational Research Review, 22*, 61–75. https://doi.org/10.1016/j.edurev.2021.100312

Popenici, S. A. D., & Kerr, S. (2017). Exploring the ethical implications of artificial intelligence in higher education. In *Proceedings of the 3rd international conference on artificial intelligence in education* (pp. 162–169). Springer.

Romero, C., & Ventura, S. (2010). Educational data mining: A review of the state of the art. *IEEE Trans. Syst. Man Cybern. Part C, 40*, 601–618.

Siemens, G., & Long, P. (2011). Penetrating the fog: Analytics in learning and education. *EDUCAUSE Rev., 46*, 30–39.

Slade, S., & Prinsloo, P. (2013). Learning analytics: Ethical issues and dilemmas. *The American Behavioral Scientist, 57*, 1510–1529.

Sukhnandan, L., & Underwood, J. (2019). Exploring the role of data analytics in the future of outcome-based education. *Education and Information Technologies, 24*, 1041–1060. https://doi.org/10.1007/s10639-018-9836-1

Tondeur, J., & van Braak, J. (2017). The role of teachers in integrating artificial intelligence into education. *Computers in Education, 115*, 88–98. https://doi.org/10.1016/j.compedu.2017.08.010

Toxigon. (2025a). AI in education: Transforming learning in 2025. *Toxigon Blog*. https://toxigon.com/ai-in-education-transforming-learning-in-2025

Toxigon. (2025b). AI in education: Top trends to watch in 2025. *Toxigon Blog*. https://toxigon.com/ai-in-education-2025-trends

Toxigon. (2025c). AI transforming education in 2025: Insights and impacts. *Toxigon Blog*. https://toxigon.com/ai-transforming-education-2025

Zhu, M., & Shen, S. (2021). Machine learning for real-time feedback in adaptive learning environments. *Journal of Educational Computing Research, 59*, 625–641. https://doi.org/10.1177/07356331211007625

Ethical Considerations and Data Privacy in ML-Driven Outcome-Based Education

B. Aruna Devi and V. Saravanan

Abstract The inclusion of machine learning (ML) into outcome-based education (OBE) has transformed academic assessment, course delivery, and the creation of individualized learning paths. Learning management systems based on machine learning enable teachers to tailor their teaching strategies to align with learning objectives, leveraging data-driven insights into student performance and the effectiveness of the curriculum. Conversely, the use of these technologies raises several significant ethical questions, particularly regarding data privacy, informed consent, algorithmic bias, and transparency in decision-making. These challenges draw attention to a substantial issue: especially when student data is aggregated or processed in centralized architectures, the current machine learning-based OBE models do not adequately guarantee data privacy, model fairness, or ethical compliance. Improper handling of this data could result in student privacy breaches or unfair evaluations. Addressing these problems builds confidence and ensures continuous GDPR and FERPA privacy rule compliance. This article presents a Privacy-Preserving Ethical Learning Model (PPELM) designed specifically for machine learning-driven OBE systems, which is proposed in the present work as a means of circumventing these constraints. Results of a comparison with centralized machine learning models indicate that PPELM performs better with less data exposure and more fairness across a wide range of student groups.

Keywords Outcome-based education · Machine learning · Data privacy · Federated learning · Ethical AI

B. Aruna Devi (✉)
Department of Electronics and Communication Engineering, Dr N.G.P Institute of Technology, Coimbatore, India
e-mail: arunadevi@drngpit.ac.in

V. Saravanan
Dambi Dollo University, Dembi Dolo, Ethiopia

D. Vetrithangam et al. (eds.), *Transforming Outcome-Based Education with Machine Learning*, https://doi.org/10.1007/978-981-95-4925-2_12

1 Introduction

Machine learning (ML) integrated into outcome-based education (OBE) systems (Ferrari et al., 2024) has initially created new opportunities for adaptive curriculum design, student performance forecasting, and individualized learning. Some significant problems develop when machine learning is applied in educational settings. Often, educational data includes private personal information, including grades, demographics, and psychological profiles, which initially draws attention since it raises serious concerns about the protection of personal information and privacy (Spaho et al., 2025). Second, biased training data or data lacking all students can lead to algorithmic injustice, which harms underrepresented student groups and compromises the equity goals of contemporary education (Slade & Prinsloo, 2013). Third, most systems currently in use are based on centralized models, which are prone to single-point failures, data leakage, and lack scalability, particularly when dealing with distributed academic institutions (Marshall et al., 2022; Choudhry et al., 2024; Reddy et al., 2024). Models such as Centralized XGBoost (XGB) and Bayesian Student Modeling (BSM) rely on the centralization of raw data, which could raise institutional concerns under data protection laws, including the General Data Protection Regulation (GDPR) and the Family Educational Rights and Privacy Act (FERPA) (Shukla & Taneja, 2024). Even the most advanced methods, such as Differentially Private SGD (DP-SGD) or Federated Averaging (FedAvg), have their limits when it comes to simultaneously balancing privacy, fairness, and model accuracy (Valli et al., 2024).

A Privacy-Preserving Ethical Learning Model (PPELM) is proposed in this work as a means of circumventing these constraints, designed explicitly for machine learning-driven OBE systems. The primary objectives are:

1. A distributed and federated learning strategy will help to forecast student performance more precisely.
2. By means of differential privacy and fairness auditing tools, ethically and responsibly apply machine learning in education.

The proposed research is novel as it combines federated learning, differential privacy, and fairness auditing—a trifold framework rarely studied together in educational environments. This clarifies the uniqueness of the labor. PPELM is a model that dynamically allocates model training across institutional nodes, ensures privacy through noise infusion, and incorporates post-hoc fairness checks to support equity. This is not like traditional centralized models or those that only preserve privacy.

The significant contributions include:

- This research develops a federated machine learning system specifically for educational institutions trying to train models on distributed student data securely.
- Training with differential privacy methods helps to lower the likelihood of data being re-identified.
- A protocol for the automated auditing of fairness will be applied to seek demographic parity and representation bias.

2 Related Works

To predict student performance and customize learning, numerous studies have investigated the application of machine learning in educational domains. Historically, academic datasets have been analyzed using methods such as decision trees and support vector machines to precisely predict at-risk students or recommend interventions (Arora et al., 2024; Taherdoost, 2023; Dhirani et al., 2023). Conversely, these models sometimes lacked scalability and suffered from overfitting due to the restricted use of feature engineering.

Its durability and interpretability (Mühlhoff, 2021) led to widespread use of the gradient-boosted decision tree algorithm known as XGBoost (XGB). Because they are typically centralized, XGB models raise privacy concerns when applied to sensitive educational records, as this requires data collection at a single location. Though they are often centralized, these models offer excellent accuracy. Originally intended to address this issue, the Differentially Private Stochastic Gradient Descent (DP-SGD) algorithm ensures that adding noise to gradients prevents the reverse-engineering of personal data contributions (Avacharmal et al., 2023) during training. Although mathematically correct, DP-SGD sometimes results in a privacy-accuracy trade-off due to the noise budget (Koshiyama et al., 2024).

A different approach, federated learning (FL), has evolved such that only model updates are shared, while data is stored on local devices. This is similar to Federated Averaging (FedAvg), which was described in (Rasool et al., 2022). The FedAvg algorithm has been effectively utilized in mobile and Internet of Things environments and is now beginning to enter the educational space. FL in the field of education still lacks empirical research, however, and ensuring equity among various student populations remains a continuous challenge (Williams, 2024).

Students' learning behaviour over time can be tracked using a probabilistic framework known as Bayesian Student Modeling, or BSM (Karimian et al., 2022). Although BSM methods can manage uncertainty and prior knowledge, they sometimes require considerable domain tuning and struggle to incorporate privacy-protecting tools (Kothinti, 2024). Moreover, Bayesian networks sometimes presume the data is uniform, which may not be the case in educational institutions distributed across different sites with varied student populations (Huang et al., 2022).

Recent research using hybrid models has attempted to combine personalisation and privacy. Hybrid differential-federated systems have been examined as a case study for the healthcare and financial sectors; education has been considered only occasionally (Khalid et al., 2023). Other projects have tried to implement fairness-aware machine learning models (Wylde et al., 2022) using in-processing bias reduction techniques and pre-processing. Though given the complexity of these models, their application in real-time systems has been limited.

Ethical machine learning auditing is another path to investigate. Though most of these audits are post-hoc and not part of model training pipelines, several studies have recommended fairness audits based on demographic parity or equal opportunity (Chy & Buadi, 2024). Most approaches to ethical compliance in machine learning for education are reactive rather than proactive, and there is very little guidance on real-time intervention (Farshad et al., 2024).

The proposed PPELM addresses these gaps by combining three fundamental pillars into a single pipeline. Fairness auditing for ethical assurance, differential privacy for secure computation, and federated learning for decentralization are these three pillars. By benchmarking against existing models like XGB, DP-SGD, FedAvg, and BSM, the model demonstrates how these technologies can be effectively coordinated to create reliable educational AI systems.

In today's learning environments, where artificial intelligence not only enhances teaching but also needs to protect student autonomy, equity, and dignity, such integration represents a significant advance over traditional approaches (Kohli et al., 2025; Sharma et al., 2024; Sikder et al., 2024).

3 Proposed Privacy-Preserving Ethical Learning Model (PPELM)

Using Federated Learning (FL), PPELM conducts distributed training across multiple institutions or nodes. Though every node is responsible for locally training a model on student data, only the updated model parameters are shared with a central aggregator; the raw data itself is not shared. Differential Privacy (DP) is applied to the shared gradients to enhance data privacy further. Several fairness criteria, including demographic parity and equal opportunity measurements, guide the ethical audit process.

3.1 Process of the Proposed Method

The proposed method begins by initializing the base model's parameters with predefined starting weights. The model is distributed to participating institutions, allowing for decentralized access to training. At each local node, the model undergoes training using institution-specific student data to capture localized patterns. To preserve privacy, controlled noise is applied to the gradients—a differential privacy measure—before sharing updates. The server then aggregates these modified gradients across all nodes, synthesizing them to update the global model. This cycle of local training, privacy adjustments, and global aggregation repeats until the model converges or reaches a predetermined epoch limit. Finally, the model undergoes a fairness audit, where standardized metrics evaluate ethical balance across demographic groups to ensure equitable outcomes.

3.2 Pseudocode

```
Initialize global_model_weights
For each round in training_rounds:
For each node in parallel:
local_data = load_student_data(node)
local_model = train_model(global_model_weights, local_data)
noisy_gradients = add_differential_privacy(local_model.gradients)
send_to_server(noisy_gradients)
global_model_weights = aggregate_updates(noisy_gradients)
evaluate_fairness(global_model_weights)
Return global_model_weights
```

3.3 Initialize Model Parameters

At the outset, a central server initializes the global model architecture and its parameters, typically weights and biases. This model is not yet trained, but it serves as the base for all local nodes to begin their training. A simple Multilayer Perceptron (MLP) with input features (e.g., student assessment scores, attendance, engagement metrics) and an output node for predicting learning outcome levels (e.g., low, medium, high) is used.

Table 1 illustrates the initialized model weights before any training begins.

The model is intentionally kept lightweight to allow for scalability and ease of deployment across distributed nodes.

3.4 Distribution of the Global Model to Local NODES

After initialization, the model is distributed to *n* participating institutions (nodes), each with its own local dataset. Each institution receives the same model structure and weights, ensuring consistent baseline training across the network. The process avoids centralizing any raw student data, supporting privacy-preserving design. Table 2 illustrates how the global model is distributed to three local nodes, each with its own subset of student data.

This design ensures every node operates independently, reducing vulnerability to data breaches.

Table 1 Initial global model weights

Layer	Weight Matrix Shape	Initial Values (Sample)
Input → hidden	(10 × 32)	0.005, −0.001, ..., 0.002
Hidden → output	(32 × 3)	−0.007, 0.003, ..., 0.001

Table 2 Distribution of initial model and data per node

Node ID	Dataset Size	Data Attributes	Received Model Version
Node 1	3200	Scores, attendance, feedback	Version 1
Node 2	3400	Scores, activity logs	Version 1
Node 3	3400	Participation, GPA, exams	Version 1

Table 3 Locally updated model gradients (differentially private)

Node ID	Gradient Noise Applied	Gradient Summary (Sample)
Node 1	Gaussian noise (σ = 1.0)	0.0023, −0.0008, ..., 0.0012
Node 2	Laplace noise (b = 0.5)	−0.0015, 0.0006, ..., 0.0003
Node 3	Gaussian noise (σ = 1.0)	0.0011, 0.0009, ..., −0.0007

3.5 Local Training

Each node performs local training using its own student data. The model is updated over several epochs, during which time gradients are computed based on local loss functions (e.g., categorical cross-entropy for multi-class outcome prediction). The updates are not shared directly; instead, *differential privacy* is applied to mask sensitive signal traces in the gradient updates.

The training at each node follows standard backpropagation, where the error between the actual and predicted learning outcomes is minimized, shown in Eq. 1:

$$\mathcal{L} = -\sum_{C}^{i=1} y_i \log\left(\hat{y}_i\right) \tag{1}$$

After training, each node sends only the *noisy* updated gradients to the central server for aggregation, ensuring privacy and ethical compliance. As illustrated in Table 3, each node generates its own set of updated (private) gradients after local training.

This step enables secure contribution to the global model without exposing any raw data or individualized patterns.

3.6 Differential Privacy

Once each node completes its local training, the updated gradients are not sent directly to the server. Instead, Differential Privacy (DP) is applied to ensure that no individual student's information can be inferred from the model updates. Each node injects calibrated noise into its gradient updates using a noise distribution (Gaussian or Laplace), thereby maintaining the privacy guarantee defined by a parameter ε (epsilon), where a lower ε indicates stronger privacy. The private gradient update g ~ \tilde{g}g ~ for each node is calculated in Eq. 2 as:

$$\tilde{g} = g + \mathcal{N}\left(0, \sigma^2\right) \tag{2}$$

where g is the original gradient and $\mathcal{N}\left(0, \sigma^2\right)$ is Gaussian noise with standard deviation σ, controlling the noise intensity. Table 4 shows the effect of DP on gradient values for three example nodes.

Applying DP ensures each node contributes useful updates to the global model while maintaining individual data privacy (see Table 4).

3.7 *Aggregate Global Model*

The central server collects the differentially private gradients from all participating nodes and performs a Federated Averaging (FedAvg) operation to update the global model. Each node's contribution may be weighted based on dataset size or training performance.

The aggregated global gradient G is calculated as a weighted average of the private gradients, provided in Eq. 3:

$$G = \frac{1}{N} \sum_{i=1}^{N} \tilde{g}_i \tag{3}$$

where $\tilde{g}_i$ represents the differentially private gradient from node i, and N is the total number of nodes. Table 5 presents an example of aggregated updates across three nodes.

The result is a globally updated model that benefits from all node inputs while maintaining decentralized privacy protections (see Table 5).

3.8 *Fairness Audit Report*

Once the updated global model is produced, it undergoes ethical auditing using fairness metrics to ensure the predictions do not disproportionately favor or disadvantage any subgroup (e.g., gender, socioeconomic status, or academic background). Metrics like Demographic Parity, Equal Opportunity, and Disparate Impact are

Table 4 Sample gradient updates after applying differential privacy

Node ID	Original Gradient (g)	Noise (σ = 1.0)	Private Gradient (g ~ \tilde{g}g ~)
Node 1	[0.01, −0.02, 0.015]	[0.003, −0.007, 0.004]	[0.013, −0.027, 0.019]
Node 2	[−0.005, 0.01, −0.008]	[0.001, −0.002, 0.003]	[−0.004, 0.008, −0.005]
Node 3	[0.02, 0.01, −0.01]	[−0.005, 0.002, 0.006]	[0.015, 0.012, −0.004]

Table 5 Aggregated global model update

Feature Index	Node 1	Node 2	Node 3	Aggregated Gradient (G)
1	0.013	−0.004	0.015	0.008
2	−0.027	0.008	0.012	−0.00233
3	0.019	−0.005	−0.004	0.00333

Table 6 Fairness audit results across demographic groups

Demographic group	Positive prediction rate	Ground truth positive rate	Disparate impact
Male	0.68	0.70	0.97
Female	0.71	0.72	0.99
Rural background	0.60	0.62	0.97
Urban background	0.74	0.76	0.97

computed to identify any predictive bias. For example, Demographic Parity ensures that the prediction rates are similar across sensitive groups, shown in Eq. 4:

$$P\left(\hat{Y}=1 \mid A=0\right) \approx P\left(\hat{Y}=1 \mid A=1\right) \tag{4}$$

Where A is a binary sensitive attribute (e.g., 0 for male and 1 for female), Table 6 displays the fairness audit results for the trained model.

All Disparate Impact values are close to 1, suggesting the model is fair across groups (see Table 6).

3.9 Results and Discussion

The model was implemented in Python using TensorFlow Federated (TFF). Simulations were conducted on a system equipped with an Intel Core i7 processor, 32 GB of RAM, and an NVIDIA RTX 3080 GPU. Experiments utilized an anonymized educational dataset from three universities, comprising 10,000 student records across eight outcome-based courses. The performance of PPELM was compared with four existing methods, including the Centralized XGBoost Model (XGB), which focuses on high accuracy but compromises data privacy. **Differentially Private SGD (DP-SGD)** is used to achieve moderate privacy at the expense of lower accuracy. **Federated Averaging (FedAvg)** focuses on decentralization, but lacks a robust privacy layer. **Bayesian Student Modeling (BSM)** focuses on strong probabilistic inference and weak scalability, gradual increase, suggesting a decrease in initial Setup and Parameters. The PPELM outperforms all four in maintaining a strong balance between accuracy, ethical fairness, and data privacy (Table 7).

Table 7 Experimental setup and parameters

Parameter	Value
Number of rounds (epochs)	50
Batch size	64
Learning rate	0.01
Privacy budget (epsilon)	1.0
Number of nodes	5 (simulated institutions)
Dataset size	10,000 records
Model type	Multilayer perceptron (MLP)
Optimizer	Adam
Evaluation Split	80% train, 20% test

3.9.1 Performance Metrics

1. **Accuracy**: The term accuracy refers to the correctness of the forecast of student outcomes.

$$Accuracy = \frac{TP + TN}{TP + TN + FP + FN}$$

2. **Privacy Leakage Risk**: This risk study aims to estimate the likelihood of reassembling private student data depending on model parameters.

$$Privacy\,Leakage\,Risks = \frac{Number\,of\,Successful\,Inferences}{Total\,Number\,of\,Inference\,Attempts}$$

3. **Fairness Index**: Assesses whether or not sensitive characteristics, such as gender, age, and academic background, are spread uniformly across the projection.

$$Fairness\,Index(FI) = 1 - Variance(Prediction\;\;rates\,across\,Groups)$$

4. **Communication Overhead**: The term "communication overhead" refers to the bandwidth and data transfer cost produced for every training round in a federated environment.

$$Communication\,Overhead(CO) = N * S$$

Where N is the Number of devices/Clients participating.

S=Size of data transmitted per client (in bytes).

PPELM reveals the lowest likelihood of privacy leakage from distributed training and integrated differential privacy. This is a significant improvement over centralized methods such as XGB and BSM, as shown in Table 8.

As shown in Table 9, PPELM's fairness index indicates a slow increase, implying less prejudice across all demographic categories. Its auditing system ensures fairness in forecast outcomes, as well as adherence to ethical standards.

The communication overhead exists in megabytes (MB) according to Table 10. The table presents detailed information about data transfer expenses that exist between various models or system configurations. System communication efficiency is assessed through this method. Optimized update compression, which preserves model fidelity, helps PPELM maintain a modest communication overhead lower than that of other federated models, such as FedAvg and DP-SGD.

As shown in Table 11, across all epochs, the proposed PPELM outperforms all baseline methods in terms of accuracy. The integration of both local learning and privacy-aware global updates helps preserve essential predictive signals, thereby enhancing performance.

Table 8 Privacy leakage risk (lower is better)

Epochs	XGB	DP-SGD	FedAvg	BSM	PPELM (Proposed)
10	0.40	0.21	0.35	0.33	0.19
20	0.38	0.19	0.31	0.30	0.17
30	0.36	0.18	0.28	0.28	0.15
40	0.34	0.17	0.26	0.26	0.13
50	0.33	0.17	0.24	0.24	0.12

Table 9 Fairness index (closer to 1 is better)

Epochs	XGB	DP-SGD	FedAvg	BSM	PPELM (Proposed)
10	0.85	0.87	0.89	0.88	0.91
20	0.86	0.89	0.91	0.90	0.93
30	0.87	0.89	0.92	0.91	0.94
40	0.88	0.90	0.93	0.92	0.96
50	0.88	0.90	0.94	0.92	0.97

Table 10 Communication overhead (MB)

Epochs	XGB	DP-SGD	FedAvg	BSM	PPELM (Proposed)
10	5.0	6.8	7.2	5.6	6.5
20	10.0	13.4	14.1	11.2	12.7
30	14.8	20.1	21.0	16.5	18.9
40	19.7	27.5	28.3	22.0	25.1
50	24.5	34.1	35.5	27.3	31.0

Table 11 Accuracy comparison (%) over epochs

Epochs	XGB	DP-SGD	FedAvg	BSM	PPELM (Proposed)
10	72.3	68.1	74.5	70.8	75.9
20	76.0	70.5	77.3	73.2	79.4
30	78.8	73.4	79.5	75.9	82.6
40	80.2	74.8	81.2	77.1	84.0
50	81.5	75.2	82.0	78.0	85.1

4 Conclusion and Future Work

The proposed Privacy-Preserving Ethical Learning Model (PPELM) promises student data confidentiality, fairness, and accurate performance forecasts, therefore representing a significant development in outcome-based education. The PPELM system combines ethical concerns, distributed intelligence, and user privacy protection technologies to aim to meet the demands of modern digital education platforms, which value trustworthiness, inclusiveness, and accuracy. Its design makes it especially suitable for scalable deployment in businesses seeking the ethical use of AI-driven insights to tailor learning while maintaining strict data governance policies. In the future, this work can be enhanced by utilizing several transfer learning mechanisms and methodologies within a real-time learning model to improve data confidentiality, fairness, and ethical considerations.

References

Arora, S., Thota, S. R., & Gupta, S., (2024). *Data mining and processing in the age of big data and artificial intelligence-issues, privacy, and ethical considerations.* In 2024 4th IEEE Asian Conference on Innovation in Technology (ASIANCON), pp. 1–6.

Avacharmal, R., Gudala, L., & Venkataramanan, S. (2023). Navigating the labyrinth: A comprehensive review of emerging artificial intelligence technologies, ethical considerations. And global governance models in the pursuit of trustworthy AI. *Australian Journal of Machine Learning Research & Applications, 3*(2), 331–347.

Choudhry, M. D., Sundarrajan, M., Jeevanandham, S., & Saravanan, V. (2024). Security and privacy issues in AI-based biometric systems. In *AI-based advancements in biometrics and its applications* (pp. 85–100).

Chy, M. K. H., & Buadi, O. N. (2024). Role of machine learning in policy making and evaluation. *International Journal of Innovative Science and Research Technology (IJISRT), 9*, 456. https://doi.org/10.38124/ijisrt/IJISRT24OCT687

Dhirani, L. L., Mukhtiar, N., Chowdhry, B. S., & Newe, T. (2023). Ethical dilemmas and privacy issues in emerging technologies: A review. *Sensors, 23*(3), 1151.

Farshad, S., Zorin, E., Amangeldiuly, N., & Fortin, C. (2024). Engagement assessment in project-based education: A machine learning approach in team chat analysis. *Education and Information Technologies, 29*(10), 13105–13131.

Ferrari, M. D., Wang, Y., Curcin, V., & Edgeworth, J. (2024). Multi-objective symbolic regression for clinician-in-the-loop in machine learning algorithms. *European Journal of Public Health, 34*, 1.

Huang, C., Zhang, Z., Mao, B., & Yao, X. (2022). An overview of artificial intelligence ethics. *IEEE Transactions on Artificial Intelligence, 4*(4), 799–819.

Karimian, G., Petelos, E., & Evers, S. M. (2022). The ethical issues of the application of artificial intelligence in healthcare: A systematic scoping review. *AI and Ethics, 2*(4), 539–551.

Khalid, N., Qayyum, A., Bilal, M., Al-Fuqaha, A., & Qadir, J. (2023). Privacy-preserving artificial intelligence in healthcare: Techniques and applications. *Computers in Biology and Medicine, 158*, 106848.

Kohli, M., Kar, A. K., Sinha, S., & Kohli, S. (2025). Stakeholder perception towards a machine-learning-based digital platform for detection and management of autism spectrum disorder. *Expert Systems, 42*(1), e13545.

Koshiyama, A., Kazim, E., Treleaven, P., Rai, P., Szpruch, L., Pavey, G., et al. (2024). Towards algorithm auditing: Managing legal, ethical and technological risks of AI, ML and associated algorithms. *Royal Society Open Science, 11*(5), 230859.

Kothinti, R. R. (2024). Artificial intelligence in healthcare: Revolutionizing precision medicine, predictive analytics, and ethical considerations in autonomous diagnostics. *World Journal of Advanced Research and Reviews, 19*(3), 3395–3406.

Marshall, R., Pardo, A., Smith, D., & Watson, T. (2022). Implementing next-generation privacy and ethics research in education technology. *British Journal of Educational Technology, 53*(4), 737–755.

Mühlhoff, R. (2021). Predictive privacy: Towards an applied ethics of data analytics. *Ethics and Information Technology, 23*(4), 675–690.

Rasool, R. U., Ahmad, H. F., Rafique, W., Qayyum, A., & Qadir, J. (2022). Security and privacy of internet of medical things: A contemporary review in the age of surveillance, botnets, and adversarial ML. *Journal of Network and Computer Applications, 201*, 103332.

Reddy, J. K., Syed, W. K., Mohammed, A., Gupta, K., & Natarajan, Y. (2024, September). Improved security analysis for e-commerce transactions using an AI-based decision framework. In *2024 international conference on artificial intelligence and emerging technology (global AI summit)* (pp. 864–869). IEEE.

Sharma, A. K., Dubey, R. K., Goswami, P., & Dandotiya, N. (2024). Transforming credit risk management with machine learning in the financial sector. In *2024 4th IEEE International Conference on Advancement in Electronics & Communication Engineering (AECE)* (pp. 1323–1328).

Shukla, R. P., & Taneja, S. (2024). Ethical considerations and data privacy in artificial intelligence. In *Integrating generative AI in education to achieve sustainable development goals* (pp. 86–97). IGI Global.

Sikder, A. S., Rifat, M. S., Akter, N., & Rahman, A. (2024). Harnessing machine learning for operational excellence: Transforming business efficiency across sectors: Machine learning for transforming business efficiency. *International Journal of Imminent Science & Technology, 2*(2), 123.

Slade, S., & Prinsloo, P. (2013). Learning analytics: Ethical issues and dilemmas. *American Behavioral Scientist, 57*(10), 1510–1529.

Spaho, E., Çiço, B., & Shabani, I. (2025). IoT integration approaches in personalized online learning: A systematic review. *Computers, 14*(2), 63.

Taherdoost, H. (2023). Navigating the ethical and privacy concerns of big data and machine learning in decision making. *Intelligent and Converged Networks, 4*(4), 280–295.

Valli, L. N., Sujatha, N., Mech, M., & Lokesh, V. S. (2024). Ethical considerations in data science: Balancing privacy and utility. *International Journal of Science and Research Archive, 11*(1), 11–22.

Williams, R. T. (2024). The ethical implications of using generative chatbots in higher education. *Frontiers in Education, 8*, 1331607.

Wylde, V., Rawindaran, N., Lawrence, J., Balasubramanian, R., Prakash, E., Jayal, A., & Platts, J. (2022). Cybersecurity, data privacy and blockchain: A review. *SN Computer Science, 3*(2), 127.

MIX
Papier aus verantwortungsvollen Quellen
Paper from responsible sources
FSC® C105338

If you have any concerns about our products,
you can contact us on
ProductSafety@springernature.com

In case Publisher is established outside the EU,
the EU authorized representative is:
Springer Nature Customer Service Center GmbH
Europaplatz 3, 69115 Heidelberg, Germany

Printed by Libri Plureos GmbH
in Hamburg, Germany